ST. MARY'S CITY, MARYLAND 20686

MATHEMATICS FROM MANUSCRIPT TO PRINT
1300–1600

Mathematics from Manuscript to Print
1300–1600

Edited by
CYNTHIA HAY
Open University

CLARENDON PRESS · OXFORD
1988

Oxford University Press, Walton Street, Oxford OX2 6DP
Oxford New York Toronto
Delhi Bombay Calcutta Madras Karachi
Petaling Jaya Singapore Hong Kong Tokyo
Nairobi Dar es Salaam Cape Town
Melbourne Auckland
and associated companies in
Berlin Ibadan

Oxford is a trade mark of Oxford University Press

Published in the United States
by Oxford University Press, New York

© Cynthia Hay, 1988

All rights reserved. No part of this publication may be reproduced, stored in a retrieval system, or transmitted, in any form or by any means, electronic, mechanical, photocopying, recording, or otherwise, without the prior permission of Oxford University Press

British Library Cataloguing in Publication Data
Mathematics from manuscript to print, 1300–1600.
1. Mathematics—History
I. Hay, Cynthia
510'.9'023   QA23
ISBN 0-19-853909-6

Library of Congress Cataloging in Publication Data
Mathematics from manuscript to print. 1300–1600
edited by Cynthia Hay.
Includes index.
1. Mathematics—History. I. Hay. Cynthia.
QA23.M28   1987   510'.9—dc19   87-16002 CIP
ISBN 0-19-853909-6

Set by Colset Private Ltd.
Printed in Great Britain by
St Edmundsbury Press,
Bury St Edmunds, Suffolk

PREFACE

THIS volume records the proceedings of a joint conference of the British Society for the History of Mathematics and the Open University on Renaissance mathematics, held at Keble College, Oxford, September 27–30, 1984. It should be said here that the idea of holding an Anglo-French conference on Renaissance mathematics in 1984, to celebrate the quincentenary of the completion of Nicolas Chuquet's *Triparty*, originated with Graham Flegg, who would have jointly edited the proceedings with me, but for an unfortunate bout of ill health. The conference proceedings, with a wide range of mathematical notation and references, were efficiently and ingeniously word-processed by Barbara Moss; her expertise in both the history of mathematics and in French has been invaluable in preparing these proceedings for publication. Several of the papers were originally written in French. M L'Huillier's paper was translated by Barbara Moss; the papers by Professor Beaujouan, M Benoit and Dr Cassinet were translated by Cynthia Hay.

Financial support for the conference came from the Royal Society, the London Mathematical Society, the Open University, the Société française d'histoire des sciences et des techniques, and the European Educational Research Trust.

Milton Keynes   C.H.
April 1987

CONTENTS

Contributors ix
Introduction 1
C. HAY

PART I: Italian and Provençal mathematics 9

Fourteenth-century Italian algebra 11
R. FRANCI AND L. TOTI RIGATELLI

On an algorithm for the approximation of surds from a Provençal treatise 30
J. SESIANO

PART II: Nicolas Chuquet and French mathematics 57

Nicolas Chuquet—an introduction 59
G. FLEGG

The place of Nicolas Chuquet in a typology of fifteenth-century French arithmetics 73
G. BEAUJOUAN

Concerning the method employed by Nicolas Chuquet for the extraction of cube roots 89
H. L'HUILLIER

The commercial arithmetic of Nicolas Chuquet 96
P. BENOIT

Chuquet's mathematical executor: could Éstienne de la Roche have changed the history of algebra? 117
B. MOSS

How algebra came to France 127
W. VAN EGMOND

PART III: Mathematics in the sixteenth century 145

What could we learn from Master Christianus van Varenbraken? 147
M. KOOL

A note on Rudolf Snellius and the early history of mathematics in Leiden 156
K. VAN BERKEL

The first arithmetic book of Francisco Maurolico, written in 1557
and printed in 1575: a step towards a theory of numbers   162
J. CASSINET

PART IV: Mathematics and its ramifications   181

Is translation betrayal in the history of mathematics?   183
C. HAY

Renaissance mathematics (and astronomy) in Baldassare
Boncompagni's *Bullettino di bibliografia e di storia delle scienze
matematiche e fisiche* (1868–87)   190
S. A. JAYAWARDENE

Some early sources in recreational mathematics   195
D. SINGMASTER

Cornelius Agrippa's mathematical magic   209
A. G. MOLLAND

The market place and games of chance in the fifteenth and sixteenth
centuries   220
I. SCHNEIDER

Perspective and the mathematicians: Alberti to Desargues   236
J. V. FIELD

Why did mathematics begin to take off in the sixteenth century?   264
G. J. WHITROW

Index   271

CONTRIBUTORS

G. BEAUJOUAN, École Pratique des Hautes Études, IVe section, Sciences historiques et philologiques, 45–47 rue des Écoles, 75005 Paris, France.

P. BENOIT, Centre National de la Recherche Scientifique, Université de Paris I, 9 rue Malher, 75004 Paris, France.

K. VAN BERKEL, Rijksuniversiteit Utrecht, Instituut voor geschiedenis der natuurwetenschappen, Janskerhof 30, 3512 BN Utrecht, Netherlands.

J. CASSINET, Université Paul Sabatier, Toulouse, France.

W. VAN EGMOND, Institut für Geschichte der Wissenschaften der Universität München, Deutsches Museum, D-8000 München 26, Federal Republic of Germany.

J. V. FIELD, Science Museum, South Kensington, London SW7 2DD, UK.

G. FLEGG, Faculty of Mathematics, The Open University, Walton Hall, Milton Keynes MK7 6AA, UK.

R. FRANCI, Università di Siena, Centro Studi della Matematica Medioevale, Via del Capitano 15, Siena 53100, Italy.

C. HAY, Faculty of Mathematics, The Open University, Walton Hall, Milton Keynes MK7 6AA, UK.

H. L'HUILLIER, Compagnie Française des Petroles, TOTAL, 5 rue Michel-Ange, 75781 Paris, France.

S. A. JAYAWARDENE, Science Museum, South Kensington, London SW7 2DD, UK.

M. KOOL, Jhr. Ranweg 28b, 3998 JR. Schalkwijk, Netherlands.

G. MOLLAND, Department of History and Philosophy of Science, King's College, University of Aberdeen, Aberdeen AB9 2UB, UK.

B. MOSS, 36 Haroldstone Road, London E17 7AW, UK.

I. SCHNEIDER, Institut für Geschichte der Wissenschaften der Universität München, Deutsches Museum, D-8000 München 26, Federal Republic of Germany.

J. SESIANO, Département des Mathématiques, École Polytechnique Fédérale de Lausanne, CH-1015, Lausanne, Switzerland.

D. SINGMASTER, Department of Mathematics, Polytechnic of the South Bank, Borough Road, London SE1 0AA, UK.

L. TOTI RIGATELLI, Università di Siena, Centro Studi della Matematica Medioevale, Via del Capitano 15, Siena 53100, Italy.

G. J. WHITROW, Department of Mathematics, Imperial College of Science and Technology, Huxley Building, Queen's Gate, London SW7 2BZ, UK.

and more distinguished of their predecessors. It is reminiscent of the stuff of many a novel, in which an upwardly mobile individual 'forgets' or ignores the lowly associates who helped along the way. The social and intellectual rise of algebra may have contributed to these manuscripts having been 'hidden from history' until quite recently. It may be time for a figure as famous in the history of mathematics as Pacioli to be reassessed in the light of these manuscripts.

One theme that recurs in a number of the papers on manuscripts is that of tracing 'priority claims'—that is, investigating whether a particular mathematical technique has been correctly attributed to its originator, if indeed its originator can be established at all. Different tactics are used in these studies. Sometimes the results are agnostic, in the sense that they demolish rather than establish such claims, and matters are left open for further investigation. M L'Huillier explores techniques for extracting cube roots, and argues that the technique attributed to Cardano is not original with him; moreover, credit for this cannot be given to Chuquet, despite his having used a similar technique. The ultimate originator is as yet unknown. It is a matter of disproving the accepted view, withdrawing credit from one of the 'hero-figures' of Renaissance mathematics, and locating the origins earlier in time than had previously been thought.

Dr Jacques Sesiano explores the extraction of surds in a Provençal manuscript of *c.* 1430, with references and comparisons that range from Aristotle to Cantor. Both he and Professor Beaujouan have independently shown, by reference to this manuscript, that the one passage in his manuscript where Chuquet appears to claim originality must be read with some care for Chuquet's meaning, which is precisely expressed. Chuquet wrote, with reference to the rule of intermediate terms, a technique for approximating roots, that 'formerly [he] had been [its] inventor'.[11] This pasage is on the face of it puzzling. There are similar approximation techniques to be found earlier, so that Chuquet cannot be said to have invented this technique in its entirety; furthermore, why should it matter that Chuquet thought he had invented this technique *formerly*? Professor Beaujouan reconstructs the intellectual genealogies of a number of fifteenth-century French manuscripts, and resolves this puzzle. He gives thereby a fascinating picture of mathematical activity in fifteenth-century France, as recorded in extant manuscripts. The technique in question, the rule of intermediate terms, is to be found in the Pamiers manuscript analysed by Dr Sesiano. Chuquet's contribution, by comparison with this manuscript, is to have refined this technique, rather than to have devised it. Professor Beaujouan argues that Chuquet worked on the rule of intermediate terms before encountering the Pamiers manuscript; on reading this manuscript, he realized that he had been less original than he

---

[11] In A. Marre's edition, cited in the papers referred to: '*Il y a aussi la rigle des nombres moyens de laquelle Jadiz je fuz Inuenteur . . .*'

had thought; he had thought *formerly* that he had invented it, but now recognized that he had only modified and improved an existing rule. That Chuquet did not mention the Pamiers manuscript specifically is in keeping with his own practice, and a more general intellectual style of the time, of not citing sources and references. As Professor Beaujouan remarks at the end of his paper, there are still discoveries to be made in the field of manuscripts.

Barbara Moss's study of the relationship of de la Roche's *Lárismethique nouellement composée* to Chuquet's *Triparty* can be considered in the context of priority claims. It has often been said that, at the very least, de la Roche took advantage of Chuquet, in that Chuquet's mathematical manuscripts passed into the hands of de la Roche and remained unpublished. De la Roche's *Arismethique* incorporated much of Chuquet's work, while omitting most of what subsequent historians have found interesting and sophisticated in the way of mathematics. What might now be called plagiarism can perhaps be seen in a different light at a time when ideas of intellectual property were not quite the same as ours. De la Roche was by no means the only mathematician to incorporate the work of others into his own publications; Vasari, in his *Lives of the artists*, accused Pacioli of having passed off parts of Piero della Francesca's mathematical work as his own.[12] Barbara Moss re-examines the relationship between the work of Chuquet and that of de la Roche, and gives an apologia for the latter.

Two papers in this section on Chuquet and French mathematics consider mathematics in broader historical settings. M Paul Benoit provides a detailed account of the contents of Chuquet's Commercial Arithmetic and considers the links between this work and commercial practices at the time; he examines as well the complex and difficult question of whether commercial activities can be said to have influenced the development of mathematical work in fields other than commercial arithmetic. Dr Warren van Egmond provides a survey and an analysis of the development of French mathematics, with the much more rapid and fruitful development of Italian mathematics as a point of reference. He argues that the reasons for the relative backwardness of French mathematics lie in the lack of an active and continuing social tradition in which an intellectual tradition of mathematics could take root. Algebra, in the form of a coherent algebraic tradition, did not come into being in France prior to Viète.

In the section on mathematics in the sixteenth century, Marjolein Kool summarizes the contents of a sixteenth-century Dutch manuscript which she has studied in great depth. Dr Klaas van Berkel examines how mathematics was incorporated into the university curriculum in Leiden. The first professor of mathematics in Leiden, Rudolf Snellius, was originally only reluctantly admitted as a teacher at the University of Leiden; some restrictions

---

[12] A. Giorgio Vasari, *Lives of the artists*, Penguin, Harmondsworth, 1965, p. 191.

were put on what he could teach, perhaps because of the practical slant of his mathematical interests. Only after practical mathematics was successfully institutionalized, in a Dutch mathematical school established in 1600, was Snellius made a full professor. A new university, Leiden was not immediately receptive to new developments in mathematics, but followed suit quite quickly when these new developments were institutionalized elsewhere, in close proximity.

Dr Jean Cassinet discusses Francisco Maurolico's *Arithmetic* in connection with the development of number theory, and in particular of mathematical induction. Although the recovery of Diophantus was considerably more significant, Maurolico's work on figurate numbers came to the attention of both Fermat and Pascal.

The papers in the final section are concerned with the ramifications of mathematics and its history. Dr Cynthia Hay reflects, somewhat ruefully, on whether the enterprise of providing a representative selection of Chuquet's mathematical manuscripts, translated into English, can only be a pale shadow of the original.

There are two surveys of work in progress, aided and abetted by computers. Mr Jayawardene provides a foretaste of his promised systematic guide to exploiting the resources of the *Bullettino di bibliografia e di storia delle scienze matematiche e fisiche* for the history of mathematics. Most historians of medieval or Renaissance mathematics will have, at some time, consulted one or two volumes of this nineteenth-century periodical with a specific topic in mind; to browse through one or two volumes is tantalizing. Mr Jayawardene's guide will make possible a systematic exploitation of its riches. Dr David Singmaster surveys recreational mathematics, with particular reference to the period of the Renaissance, in connection with an extensive bibliography that he is preparing in this field, and a series in recreational mathematics that he is editing for Oxford University Press.

Dr George Molland looks at a little-known and perhaps 'disreputable' aspect of the literature on magic in this period. Dr Molland examines an aspect of the Renaissance interest in magic which had considerable acceptance, and which has subsequently been neglected: mathematical magic, as illustrated in the sixteenth-century work *De occulta philosophia* by Henry Cornelius Agrippa. Among the topics discussed are Agrippa's numerology, his views on magic squares, and on the abstract virtues. Molland assesses, sceptically, claims that aspects of Agrippa's mathematical magic can helpfully be viewed in structuralist terms or as an anticipation of modern group theory. He suggests, perhaps not too seriously, that ultimately Agrippa may not have had the courage of his convictions in that he did not develop his ideas on mathematical magic to their limits.

Professor Ivo Schneider makes a forceful case that the solution of the central problem in probability theory, the problem of points, was not simply a matter of pure mathematical invention or discovery. An analysis of one of the Italian manuscripts discussed above indicates that a correct algebraic solution had been found as early as the fourteenth century. Discussions of games of chance in the fifteenth and sixteenth centuries did not reach the standards of mathematical rigour displayed in this early solution, nor did they show any knowledge of its approach. This relapse in mathematical thinking about the solution of the problem of points is, Professor Schneider argues, closely linked with religious ideology and the development of the economy.

Dr J. V. Field examines the question of the relationship between perspective in the works of Renaissance artists and the mathematical understanding of perspective. She argues that, with the exception of Piero della Francesca, artists only had recourse to the mathematical rules of perspective at a 'rudimentary' level. The importance of the mathematical theory of perspective is to be found, rather, in its contribution to later developments in mathematics, which Dr Field traces.

The volume ends, appropriately, with Professor Whitrow's paper which raises a major question about the development of mathematics. The phrasing of the question, 'Why did mathematics begin to take off in the sixteenth century?', suggests analogies with W. W. Rostow's classic analysis of the Industrial Revolution, in *The stages of economic growth*,[13] which endeavoured to identify the prerequisites for the English economy to 'take off' in the late eighteenth century, into sustained economic growth. The analogy with economic development is a useful heuristic device which directs attention to the preconditions for sustained mathematical development. Professor Whitrow reviews a number of the factors which influenced the development of mathematics, and concludes with his question clarified but still unsolved.

The papers in this volume contribute to our understanding of Renaissance mathematics, both in terms of its technical contents and achievements, and also in terms of its ramifications for culture and society more generally. They show how much has been discovered, recovered and analysed in detailing the transition of mathematics from manuscript to print.

[13] Cambridge University Press, Cambridge, 1960.

PART I

Italian and Provençal mathematics

Fourteenth-century Italian algebra

R. FRANCI and L. TOTI RIGATELLI

THE significance of sixteenth century Italian algebra is due, not only to the quality of the results themselves, but also to the fact that they appeared suddenly and almost without justification: that is, they seem to be manifestations of genius independent of the study of previous results.

The only precedents usually recalled are the *Liber abaci* (1202) of Leonardo Pisano and the *Summa de arithmetica* (1494) of Luca Pacioli, and they seem negligible by comparison with sixteenth-century algebra texts.

Only recently has publication of some fourteenth- and fifteenth-century manuscripts shown a continuity of development with the earlier authors mentioned above.

The conclusions which we have drawn as to the high level reached by fourteenth-century algebra are based on a systematic study of eighteen manuscripts.

Among the more characteristic elements which we have found are the broad treatment of algebraic calculations and the presentation of the solutions of a large number of types of equations. In particular, concerning this latter subject, attempts to solve the cubic equation must be emphasized, even if the general solution was not found. Nevertheless, many noteworthy partial results were obtained, and some of them were used by Scipione dal Ferro and Girolamo Cardano. Fourteenth-century algebraists also turned their attention to the solution of some types of fourth degree equations and to equations of the form $ax^{2n} + bx^n + c = 0$. Finally, concerning the application of algebra to the solution of problems, we can observe that it is always used more to solve business problems, even though the fourteenth-century algebraists did not neglect the elaboration of new and difficult theoretical problems such as those which Leonardo Pisano presented in the fifteenth chapter of *Liber abaci*.

1. The sources

The primary sources for reconstructing the history of fourteenth century Italian algebra are about fifteen manuscripts which contain treatises on algebra.

Further, we must consider, as secondary sources, three very important manuscripts of the following century. These texts, which are compendia of the mathematical knowledge taught in abbacus schools, contain large excerpts from treatises of which the originals are missing and also many references to authors who would otherwise have been unknown.

The algebra treatises of the Italian Middle Ages and early Renaissance are usually a chapter of a larger text called a *Trattato d'abaco*.

The number of pages and the topics covered in the algebra vary; typical contents are, however, as follows:

- calculation with radicals
- calculation with algebraic monomials and polynomials
- rules for solving algebraic equations
- problems solved by algebra

The fourteenth-century manuscripts which we have examined are the following:[1]

Magl.Cl.XI,87 (1328, d). Firenze, Biblioteca Nazionale.[2]
Paolo Gerardi, *Libro di ragioni*, cc. 1^r–70^r.
'Le regole dela cosa', cc. 63^r–70^r.

Ms.1754 (1330, c). Lucca, Biblioteca Statale.[3]
Anon, *Libro di molte ragioni d'abaco*, cc. 1^r–87^r.
'Reghola della chosa', cc. 50^r–52^r.
'Delle radicie', cc. 80^r–80^v.
'Algebra', cc. 81^r–81^v.

Ricc.2252 (fourteenth century). Firenze, Biblioteca Riccardiana.
(i) Anon, *Tractato dell'aritmetica* (1350, c), cc. 1^r–71^r.
'Regole della cosa', cc. 25^r–30^r.
(ii) Anon, *Regole dell'algebre* (1400, c), cc. 160^r–169^r.

Ricc.2263 (fourteenth century). Firenze, Biblioteca Riccardiana.
Anon, *Trattato dell'alcibra amuchabile* (1365, c), cc. 24^r–50^v.

---

[1] For each listed manuscript we have shown the chapters devoted to algebra, also marking the folio numbers next to the titles. When the date quoted after the signature is followed by 'd', it means that the text is dated. The letter 'c' after the date means that the date is approximate and was determined by a combination of several parameters: watermark, paleographic elements, internal evidence, etc. All the listed manuscripts are described in W. Van Egmond (1980), *Practical mathematics in the Italian Renaissance. A catalogue of Italian abbacus manuscripts and printed books to 1600*, Firenze, Istituto e Museo di Storia della Scienza.

[2] A description and an index of this manuscript are given in G. Arrighi (1967), 'Due trattati di Paolo Gherardi matematico fiorentino. I codici Magliabechiani Cl.XI nn. 87 e 88 (prima metà del Trecento) della Biblioteca Nazionale di Firenze', *Accademia delle Scienze*, Torino, pp. 61-82. The chapter on algebra is transcribed in W. van Egmond (1978), 'The *Libro di ragioni* of Paolo Gerardi (1328)', *Physis*, 20, pp. 155-189.

[3] This manuscript has been completely transcribed and published as Scuola Lucchese (1973), *Libro d'abaco. Dal codice 1754 (sec. XIV) della Biblioteca Statale di Lucca. A cura e con introduzione di G. Arrighi*. Lucca, Cassa di Risparmio di Lucca.

Plut.30,26 (1370, d). Firenze, Biblioteca Mediceo Laurenziana.
Giovanni de' Danti. *Tractato dell'algorismo*, cc. 1^r–28^r.
'Per regola del'argibra e per muodo de propositione', cc. 20^v–21^v

L.IX.28 (1384, d). Siena, Biblioteca Comunale.[4]
Gilio, [*Aritmetica e geometria*].
'Trattato dele radici', cc. 23^r–37^r.
'Regole della cosa', cc. 37^r–45^r.

Fond.prin.II.III.198 (fourteenth century). Firenze, Biblioteca Nazionale.
(i) Anon, *Libro d'insegnare arismetricha* (1390, d), cc. 3^r–59^v.
'Delle regole delle cose', cc. 47^r–55^r.
(ii) *Liber de algiebra e almuchabila* (1390, c), cc. 86^r–107^v.

Fond.prin.II.V.152 (1390, c). Firenze, Biblioteca Nazionale.
Anon, *Tratato sopra l'arte dell'arismetrica*, cc. 1^r–180^v.
[Problemi risolti con l'algebra], cc. 109^r–144^v.
[Calcolo algebrico], cc. 145^r–153^r.
'Le reghole della cosa', cc. 153^v–166^r.
[Problemi risolti con l'algebra], cc. 169^r–180^r.

Conv.Sopp.G.7.1137 (1395, c). Firenze, Biblioteca Nazionale.
Anon, *Libro delle ragioni d'abaco*, cc. 1^r–253^r.
[Algebra], cc. 107^r–109^v.
'Reghole dell'alcibra mochabile', cc. 144^r–151^v.

Chigi M.VIII.170 (1395, c). Vaticano, Biblioteca Apostolica.[5]
Anon, *Aliabra argibra*, cc. 1^r–112^r.

Magl.Cl.XI,120 (1400, c). Firenze, Biblioteca Nazionale.
(i) Anon, [Capitoli d'algebra], cc. 1^r–7^r.
(ii) M° Antonio da Firenze, 'Regole del arzibra', cc. 7^v–10^v.
(iii) Anon, [Algebra], cc. 13^r–30^v.

---

[4] The contents of this text are described in detail in R. Franci (1983), 'Gilio da Siena: un maestro d'abaco del XIV secolo', *Atti del Convegno 'La storia delle matematiche in Italia', Cagliari, 29, 30 settembre e 1 ottobre 1982*, Bologna, pp. 319-324. The part devoted to algebra is transcribed in Gilio (1983), *Questioni d'algebra. Dal codice L.IX.28 della Biblioteca Comunale di Siena*. Quaderni del Centro Studi della Matematica Medievale, n.6, Siena.

[5] Of this important text, we also have two copies from the fifteenth century. These are **Siena I.VIII.17** (1470, c.) and **Ashburn 1199** (1495, c.). It is described in W. van Egmond (1983), 'The Algebra of Master Dardi of Pisa', *Historia mathematica*, *10*, pp. 399-421. On the basis of a report contained in the manuscript **Heb.1029** (Paris, Bibliothèque Nationale), van Egmond asserts that *Aliabra Argibra* was written by Master Dardi of Pisa in 1344. The text of *Aliabra argibra* contained in **I.VII.17** was completely transcribed and studied by three students of the University of Siena: S. Vinciani, C. Lima and M. Marini.

14 *Italian and Provençal mathematics*

L.IV.21 (1463, d). Siena, Biblioteca Comunale.[6]
M° Benedetto, *Trattato di praticha d'arismetricha*,cc. 1^r–506.
'Chasi exenplari alla reghola dell'algibra secondo che scrive Maestro Biagio', cc. 388^v–407^r.
'Chasi scritti nel trattato di Maestro Antonio nominato trattato di Fioretti', cc. 451^r–474^v.

Pal.573 (1460, c). Firenze, Biblioteca Nazionale.[7]
Anon (allievo di Domenicho d'Agostino vaiaio), *Trattato di praticha d'arismetricha*, cc. 1^r–491^r.
'El sechondo chapitolo della diecima parte che chontiene el modo di multiplichare, partire, ragugnere e trarre e detti nomi', cc. 399^r–402^r.
'Ragioni absolute dal perfectissimo arismetricho Maestro Antonio', cc. 478^v–489^v.

Ott.Lat.3307 (1465, c). Vaticano, Biblioteca Apostolica.[8]
Anon (allievo di Domenicho d'Aghostino vaiaio), *Trattato di praticha d'arismetricha*, cc. 1^r–349^v.
'Quistione sottile asolute per M° Antonio', cc.335^r–343^r.

2. Language and algebraic calculation

Most of the manuscripts which we have examined were written in the Tuscan vernacular.

---

[6] A careful description of this manuscript is given in G. Arrighi (1965), 'Il codice L.IV.21. della Biblioteca degl' Intronati di Siena e la "Bottega d'abaco" a Santa Trinita in Firenze', *Physis*, 7, pp. 369-400. This paper also contains a detailed index and the transcription of the introduction to each book and chapter. The importance of this manuscript for the history of medieval algebra is examined in R. Franci and L. Toti Rigatelli (1983), 'Maestro Benedetto da Firenze e la storia dell' algebra', *Historia mathematica,10*, pp. 297-317. All the parts devoted to algebra have been published as follows:
 Benedetto da Firenze (1982), *La reghola de algebra amuchabale. Dal codice L.IV.21*, Quaderni del Centro Studi della Matematica Medioevale n. 1, Siena.
 Leonardo Pisano (1984), *E' chasi. Dalla terza parte del XV capitolo del Liber abaci nella trascelta a cura di Maestro Benedetto*, Quaderni del Centro Studi della Matematica Medioevale n. 10, Siena.
 M° Biagio (1983), *Chasi exenplari alla reghola dell' algibra. Nella trascelta a cura di M° Benedetto*, Quaderni del Centro Studi della Matematica Medioevale n. 5, Siena.
 Giovanni di Bartolo (1982), *Certi chasi. Nella trascelta a cura di Maestro Benedetto*, Quaderni del Centro Studi della Matematica Medioevale n. 3, Siena.
 Antonio de' Mazzinghi (1967), *Trattato di Fioretti. Nella trascelta a cura di M° Benedetto*, Pisa, Domus Galileana.

[7] A very complete description of this manuscript can be found in G. Arrighi (1967), 'Nuovi contributi per la storia della matematica in Firenze nell' età di mezzo. Il codice Palatino 573 della Biblioteca Nazionale di Firenze', *Rendiconti Istituto Lombardo, Classe di Scienze (A), 101*, pp. 395-437. In this paper the introductions to each book and chapter are also transcribed.

[8] An index of this manuscript can be found in G. Arrighi (1968), 'La matematica a Firenze nel Rinascimento. Il codice Ottobobiano Latino 3307 della Biblioteca Apostolica vaticana', *Physis, 10*, pp. 70-82.

Fourteenth-century Italian algebra

Though after all the equation (1) can be easily reduced to a binomial equation, it is remarkable that formula (2) involves all the coefficients of the equation.

The formulae of group (d) concern the equations of the form:

(3) $ax^3 + bx^2 = c$
(4) $ax^3 = bx^2 + c$
(5) $ax^3 + c = bx^2$

They are present in manuscript **Fond.prin.II.V.152**, while in **Conv.Sopp.G.7.1137** we find only those relating to (3) and (5).

The solution of an equation of the form (3) is given by

(6) $x = y - \dfrac{b}{3}$

where y is a real and positive solution of the equation

(7) $y^3 = 3\left(\dfrac{b}{3}\right)^2 y + \left[c - 2\left(\dfrac{b}{3}\right)^3\right]$

By obvious changes of signs the solutions of the equations (4) and (5) can be obtained. However we must remark that the fourteenth-century abbacists did not know a formula for the solution of equations of type (7); thus in the examples presented in the above-mentioned manuscripts it is solved by trial and error.

The first definition given by the author of **Fond.prin.II.V.152** in expounding this matter is that of 'cubic root with a supplement'.[20]

In modern terminology, the cubic root of p with supplement of q is nothing but a solution of the equation $x^3 = qx + p$. In the manuscript in question the definition is presented, giving the example of the cubic root of 44 with a supplement of 5. This root is 4 because 4 satisfies the equation $x^3 = 5x + 44$.

Because, as we have already said, the calculation of cubic roots with supplement is made by trial and error, the abbacist warns that in many cases such a root can easily be found, but in others it may be impossible to calculate. Thus the y of formula (6) is the cubic root of

$$\left[c - 2\left(\dfrac{b}{3}\right)^3\right]$$

with the supplement of

$$3\left(\dfrac{b}{3}\right)^2$$

The rule given by the anonymous author of **Fond.prin.II.V.152** for equations of the form (3) is:

---

[20] 'Radice chubicha chon l'aghugnimento d'alcuno numero' (c.164ᵛ).

22 Italian and Provençal mathematics

When the *cubi* and *censi* are equal to the number, at first divide by the *cubi* and then divide the *censi* into three parts, and cube one part and add the result to the number, then consider another part of the *censi* and multiply it by itself and multiply this result by 3, and multiply the last result by the other part of the *censi* and subtract the result from the number. And the cubic root of the remainder with the supplement of what came from the second part of the *censi* when they were multiplied by themselves and afterwards by 3, minus a part of the *censi* as much as you will say that is the value of the *cosa*.[21]

The solution of the above-mentioned three cases summarized in one equation of the form:

$$x^3 + px^2 + q = 0$$

is reduced to that of the equation

(8) $\quad y^3 + p'y + q' = 0$

by the transformation $x = y - \frac{p}{3}$.[22]

This is a clear enough explanation of why, one and a half centuries later, Scipione dal Ferro limited his research to the solution of equations of type (8).

What we have said so far shows how great an interest fourteenth-century abbacists had in the solution of cubic equations. This problem was also discussed outside the abbacus schools.

In manuscript **Pal. 573**, in fact, it is told that Giovanni de' Bicci de' Medici wrote in 1396 to Masolo da Perugia, who was then teaching in Venice, to ask him for explanations of some problems. In his answer, Masolo communicates to Medici, among other things, the knowledge of how to solve some equations of the third degree. We read in the letter:

I understand that it seems important that *chubi*, *censi*, and *cose* can be equalized to the number, because the opinion of all the ancient abbacists was that it was not possible. I say to you that each equation is possible to define and, if the way were not so long and difficult, I would send it to you in this letter, however I send you what you asked me.[23]

[21] '*Quando gli chubi e censi sono iguali al numero, prima si parta ne' chubi, e poi si sterzino i censi, e l'una parte si chubichi, e ciò che fa s'agiunga al numero, e poi si pigli un'altra parte di censi e quella si multiplicha per sé medesima, e ciò che fa si multiplicha per 3, e ciò che fa si multiplicha chontro all'altra parte di censi, e quello che fa s'abatta del numero, e la radice chubicha del rimanente chollo agiugnimento che naque della seconda parte de' censi quando si multiplichò per sé medesimo e poi per 3, meno l'una parte de' censi, chotanto dirai che vaglia la chosa.*' **Fond.prin.II.III.152**, c. 164^v.

[22] Further details on these cases can be found in R.Franci (1984), 'Contributi alla risoluzione dell'equazione di 3° grado nel XIV secolo', *Mathemata. Festschrift für Prof. Dr. H. Gericke*, Wiesbaden, F. Steiner Verlag.

[23] '*Chonprendo vi paia gran fatto che chubi censi et chose si possono aguagliare al numero, chonsiderato l'opinione di tutti e' passati è suta non essere possibile, io ve dicho che onne adeguaglimento è possibile a definire et sel modo non fusse molto lungho et dificile, in questa vel manderia; bene che quello che ne richiedete, vi mando.*' **Pal. 573**,c.409^v.

We do not know what solutions were given by Masolo, but it is interesting that the questions posed by Giovanni de' Medici to Masolo concerned cubic roots with a supplement.

3.3 Fourth-degree equations

Though at least one example of a fourth-degree equation is present in each of the listed manuscripts, the treatment of this subject is very different from one text to another. Again, to simplify our analysis, we have classified formulae for solution into groups:

  α Formulae for binomial equations or equations reducible to a lower degree.
  β Formulae for equations of the type $ax^4 + bx^2c = 0$
  γ Exact formulae for the solution of two classes of complete equations.

The binomial equation $ax^4 = b$ is present in all the texts, and it is the only one of the fourth degree in manuscripts **L.IX.28** and **Magl.Cl.XI, 120 (ii)**; it is solved by two extractions of the square root.

In the texts of the first half of the century, apart from the above-mentioned binomial equations, we find only equations of the following types: $ax^4 = bx$, $ax^4 = bx^2$, $ax^4 = bx^3$.

Beginning with the manuscript **Ricc. 2263** (1365, c), we find also the cases: $ax^4 + bx^3 = cx^2$, $ax^4 = bx^3 + cx^2$, $ax^4 + cx^2 = bx^3$. The formulae for solution in question are those for quadratic equations; obviously zero solutions are ignored.

The solution of some problems of the fifteenth chapter of Leonardo Pisano's *Liber abaci* is equivalent to that of equations of the form $ax^4 + bx^2 + c = 0$, which are correctly solved even though the author does not list them among the rules.

We must wait until the last ten years of the fourteenth century to find a theoretical treatment of equations of this kind. The cases

(9) $ax^4 = bx^2 + c$
$ax^4 + bx^2 = c$
$ax^4 + c = bx^2$

are in fact present in manuscripts **Fond.prin.II.V.152**, **Conv.Sopp. G.7.1137**, **Chigi M.VIII.170**, **Ricc.2252 (ii)**, **Magl.Cl.XI, 120 (iii)**. Each equation is at first solved as a quadratic equation, that is, as if it were $x^4 = y^3$; an extraction of the square root of the value found for y gives the value for x.

The treatment of this subject is particularly full in manuscript **Chigi M.VIII.170**, where, besides the above-mentioned cases, we can also find the following ones, which we list in the order and with the numbering used by the author. Beside each equation, we write the formula for solution given in the text; of course we use modern symbolism.

43. $ax^4 + bx^4 = c$ $\qquad x = \sqrt{\sqrt{\dfrac{c}{a} + \dfrac{b}{4a^2}} - \sqrt{\dfrac{b}{4a^2}}}$

75. $ax^4 + bx^2 = \sqrt{c}$ $\qquad x = \sqrt{\sqrt{\left(\dfrac{b}{2a}\right)^2 + \dfrac{\sqrt{c}}{a}} - \dfrac{b}{2a}}$

76. $ax^4 + \sqrt{c} = bx^2$ $\qquad x = \sqrt{\dfrac{b}{2a} \pm \sqrt{\left(\dfrac{b}{2a}\right)^2 - \dfrac{\sqrt{c}}{a}}}$

77. $ax^4 = bx^2 + \sqrt{c}$ $\qquad x = \sqrt{\sqrt{\left(\dfrac{b}{2a}\right)^2 + \dfrac{\sqrt{c}}{a}} + \dfrac{b}{2a}}$

118. $ax^4 + \sqrt{bx^4} = \sqrt{c}$ $\qquad x = \sqrt{\sqrt{\dfrac{b}{4a^2} + \sqrt{\dfrac{c}{a^2}}} - \sqrt{\dfrac{b}{4a^2}}}$

119. $ax^4 + \sqrt{c} = \sqrt{bx^4}$ $\qquad x = \sqrt{\sqrt{\dfrac{b}{4a^2}} \pm \sqrt{\dfrac{b}{4a^2} - \sqrt{\dfrac{c}{4a^2}}}}$

120. $ax^4 = \sqrt{bx^4} + \sqrt{c}$ $\qquad x = \sqrt{\sqrt{\dfrac{b}{4a^2}} + \sqrt{\dfrac{c}{a^2}} + \sqrt{\dfrac{b}{4a^2}}}$

130. $ax^4 + \sqrt{bx^4} = \sqrt[3]{c}$ $\qquad x = \sqrt{\sqrt{\dfrac{b}{4a^2} + \sqrt[3]{\dfrac{c}{a^3}}} - \sqrt{\dfrac{b}{4a^2}}}$

131. $ax^4 + \sqrt[3]{c} = \sqrt{bx^4}$ $\qquad x = \sqrt{\sqrt{\dfrac{b}{4a^2}} \pm \sqrt{\dfrac{b}{4a^2} - \sqrt[3]{\dfrac{c}{a^3}}}}$

132. $ax^4 = \sqrt{bx^4} + \sqrt[3]{c}$ $\qquad x = \sqrt{\sqrt{\dfrac{b}{4a^2} + \sqrt[3]{\dfrac{c}{a^3}}} + \sqrt{\dfrac{b}{4a^2}}}$

The rule applied in each of the above listed cases is clearly that for equations of type (9). The cases in question probably are presented separately because some coefficients in the equations are radicals.

The formulae of group (ϑ) are of great interest. The first of them is valid for the class of equations of the form

(10) $ax^4 + bx^3 + cx^2 + dx = e$

where

$$\dfrac{b}{a} = 4\sqrt{\dfrac{d}{b}} \text{ and } \dfrac{c}{a} = 6\dfrac{d}{b}$$

The formula for solution suggested by the fourteenth century authors is

$$x = \sqrt[4]{\left(\dfrac{d}{b}\right)^2 + \dfrac{e}{a}} - \sqrt{\dfrac{d}{b}}$$

It can be found in manuscripts **Fond.prin.II.III.198**, **Chigi M.VIII.170**, and **Ricc. 2252 (ii)**. In the first of these, the rule is presented without any explanatory problem; in the other manuscripts, the example given is the following one:

A man loaned 100 *lire* to another and after four years he gives him for the principal and the interest 160 *lire* at annual compound interest. I ask you at what rate the *lira* was loaned per month.[24]

The equation that solves the problem is

$$x^4 + 80x^3 + 2400x^2 + 32\,000\,x = 96\,000$$

whose coefficients clearly satisfy the conditions of the problem. Again, only the author of manuscript **Chigi M.VIII.170** notes that this rule is not general, and limits its validity to the class of equations that arise from the solution of four-year interest problems. Keeping the symbolism used in the similar case of the third degree, this class can be represented by the equation

$$x^4 + 80x^3 + 2400x^2 + 32\,000\,x = 16\,000\,\left(\frac{B}{A} - 1\right)$$

Again, formula (11) actually applies to a wider class in which the relations between the coefficients allow one to complete the first member of (10) to a fourth powder. Adding $\left(\dfrac{d}{b}\right)^2$

to both members of (10) we obtain

$$\left(x + \sqrt{\frac{d}{b}}\right)^4 = \frac{e}{a} + \left(\frac{d}{b}\right)^2$$

from which the formula for solution (11) follows.
The formula

$$x = \sqrt[4]{\left(\frac{c}{4a}\right)^2 + \frac{e}{a} + \frac{b}{4a}} - \sqrt{\frac{d}{2b}}$$

solves the class of the fourth-degree equations obtained from the problem of dividing 10 into two parts, namely x and *10-x*, such that

$$\frac{10x - x^2}{2x - 10} = \sqrt{N}$$

According to whether $N > / < 25$, the equation is respectively of the form:

$$ax^4 + dx = bx^3 + cx^2 + e$$
$$ax^4 + dx = bx^3 + e$$
$$ax^4 + cx^2 + dx = bx^3 + e$$

---

[24] 'Uno impresta a un altro lire 100 e in la fine de 4 anni ello riceve lire 160 intro vadagno et cave dal fazando ogn'ano cavo d'anno. Adomando a che rason fo imprestada la lira al mese?' **Chigi M.VIII.170**, c.102ᵛ.

In the **Chigi M.VIII.170** manuscript two examples are presented: in the first, N = 18, in the second, N = 28. In manuscript **Ricc. 2252 (ii)** we find only the first example.

3.4 Higher-degree equations

Equations of a degree higher than four can be found only in two manuscripts: **Chigi M.VIII.170** and **Magl.Cl.XI, 120 (ii)**. In the latter we find among the simple cases the binomial equations $ax^5 = b, ax^6 = b$. No numerical example is given in this text.

The treatment in the **Chigi M.VIII.170** manuscript is richer and more interesting. Here, some cases of equations of a degree apparently not higher than four but containing either irrational coefficients or non-integer powers of the unknown are presented. By performing algebraic calculations that permit the elimination of irrational quantities or make the degree of each coefficient integral, the equation is transformed into another equation, of degree greater than four.

We list the equations in the order and with the numbering of the author, adding to each of them the corresponding higher-degree equation.

21. $ax^3 = \sqrt{b}$ $a^2x^6 = b$
23. $ax^4 = \sqrt{b}$ $a^2x^8 = b$
31. $ax^3 = \sqrt{bx^2}$ $a^2x^6 = bx^2$
32. $ax^4 = \sqrt{bx^2}$ $a^2x^8 = bx^2$
33. $ax^3 = \sqrt{bx^3}$ $a^2x^6 = bx^3$
34. $ax^3 = \sqrt{bx^4}$ $a^2x^6 = bx^4$
35. $ax^4 = \sqrt{bx^4}$ $a^2x^8 = bx^4$
49. $ax^2 = \sqrt[3]{b}$ $a^3x^6 = b$
51. $ax^3 = \sqrt[3]{b}$ $a^3x^9 = b$
53. $ax^4 = \sqrt[3]{b}$ $a^3x^{12} = b$
57. $ax^3 = \sqrt[3]{bx}$ $a^3x^9 = bx$
62. $ax^3 = \sqrt[3]{bx^3}$ $a^3x^9 = bx^3$
63. $ax^4 = \sqrt[3]{bx^3}$ $a^3x^{12} = bx^3$
64. $ax^4 = \sqrt[3]{bx^4}$ $a^3x^{12} = bx^4$
91. $ax^3 = \sqrt{bx^3} + c$ $a^2x^6 + c^2 = (b + 2ac)x^3$
92. $ax^3 + c = \sqrt{bx^3}$ $a^2x^6 + c^2 = (b - 2ac)x^3$
93. $ax^4 + c = \sqrt{bx^4}$ $a^2x^8 + c^2 = (b - 2ac)x^4$
105. $ax^3 + b = \sqrt{c}$ $a^2x^6 + 2abx^3 = c - b^2$

Fourteenth-century Italian algebra   27

106. $ax^3 + \sqrt{b} = c$ $a^2x^6 + (c^2 - b) = 2acx^3$
107. $ax^4 + b = \sqrt{c}$ $a^2x^8 + 2abx^4 = c - b^2$
108. $ax^4 + \sqrt{b} = c$ $a^2x^8 + (c^2 - b) = 2acx^4$
165. $\sqrt[3]{ax^3} = \sqrt{b}$ $a^2x^6 = b^3$
166. $\sqrt[3]{ax^4} = \sqrt{b}$ $a^2x^8 = b^3$
168. $\sqrt{ax^2} = \sqrt[3]{b}$ $a^3x^6 = b^2$
169. $\sqrt{ax^3} = \sqrt[3]{b}$ $a^3x^9 = b^2$
170. $\sqrt{ax^4} = \sqrt[3]{b}$ $a^3x^{12} = b^2$
172. $\sqrt{ax^2} = \sqrt[3]{bx^2}$ $a^3x^6 = b^2x^4$
173. $\sqrt{ax^3} = \sqrt[3]{bx^3}$ $a^3x^9 = b^2x^6$
174. $\sqrt{ax^4} = \sqrt[3]{bx^4}$ $a^3x^{12} = b^2x^8$
176. $\sqrt{ax} = \sqrt[3]{bx^3}$ $a^3x^3 = b^2x^6$
177. $\sqrt{ax} = \sqrt[3]{bx^4}$ $a^3x^3 = b^2x^8$
178. $\sqrt{ax^2} = \sqrt[3]{bx}$ $a^3x^6 = b^2x^2$
179. $\sqrt{ax^3} = \sqrt[3]{bx}$ $a^3x^9 = b^2x^2$
180. $\sqrt{ax^4} = \sqrt[3]{bx^2}$ $a^3x^{12} = b^2x^4$
181. $\sqrt{ax^3} = \sqrt[3]{bx^4}$ $a^3x^9 = b^2x^8$
182. $\sqrt{ax^4} = \sqrt[3]{bx^3}$ $a^3x^{12} = b^2x^6$

In trinomial equations of this kind, where either a term with non-integer degree or an irrational coefficient is present, passing to the corresponding higher-degree equation would require too many calculations; thus the abbacists resort to a transformation. We list the cases in question, giving also the formula for solution suggested in the text.

41. $ax^3 + \sqrt{bx^3} = c$ $x = \sqrt[3]{\left(\sqrt{\dfrac{c}{a} + \dfrac{b}{4a^2}} - \sqrt{\dfrac{b}{4a^2}}\right)^2}$

115. $ax^3 + \sqrt{bx^3} = \sqrt{c}$ $x = \sqrt[3]{\left(\sqrt{\dfrac{b}{4a^2} + \sqrt{\dfrac{c}{a^2}}} - \sqrt{\dfrac{b}{4a^2}}\right)^2}$

116. $ax^3 + \sqrt{c} = \sqrt{bx^3}$ $x = \sqrt[3]{\left(\sqrt{\dfrac{b}{4a^2}} \pm \sqrt{\dfrac{b}{4a^2} - \sqrt{\dfrac{c}{a^2}}}\right)^2}$

117. $\sqrt{bx^3} + \sqrt{c} = ax^3$ $x = \sqrt[3]{\left(\sqrt{\dfrac{b}{4a^2} + \sqrt{\dfrac{c}{a^2}}} + \sqrt{\dfrac{b}{4a^2}}\right)^2}$

127. $ax^3 + \sqrt{bx^3} = \sqrt[3]{c}$ $x = \sqrt[3]{\left(\sqrt{\dfrac{b}{4a^2} + \sqrt[3]{\dfrac{c}{a^3}}} - \sqrt{\dfrac{b}{4a^2}}\right)^2}$

28 *Italian and Provençal mathematics*

128. $ax^3 + \sqrt[3]{c} = \sqrt{bx^3}$ $\qquad x = \sqrt[3]{\left(\sqrt{\dfrac{b}{4a^2}} \pm \sqrt{\dfrac{b}{4a^2} - \sqrt[3]{\dfrac{c}{a^3}}}\right)^2}$

129. $ax^3 = \sqrt{bx^3} + \sqrt[3]{c}$ $\qquad x = \sqrt[3]{\left(\sqrt{\dfrac{b}{4a^2}} + \sqrt[3]{\dfrac{c}{a^3}} + \sqrt{\dfrac{b}{4a^2}}\right)^2}$

Each equation is at first solved as a quadratic equation, that is, as if it were $x^3 = y^2$. The solution found is then squared, and finally a cubic root is calculated.

4. The problems

For abbacists in the Middle Ages and the early Renaissance, knowledge of algebra was a matter of professional pride; however, it was also necessary for the importance of this knowledge to be recognized by the merchants, who were the first users of the abbacists' work. It was thus necessary to show the merchants themselves the 'usefulness' of algebra.

Thus as early as the beginning of the fourteenth century we find the first applications of algebra to business problems without neglecting theoretical and speculative problems. Paolo Gerardi already illustrates some of his rules with easy business problems. M° Benedetto of Florence, author of the well-known manuscript **L.IV.21**, identifies the Florentine M° Biagio, who taught the better-known Paolo dell' Abbaco and died about 1340, as the first abbacist to reduce algebra to a 'good practice', that is, to make good use of algebra as a tool for solving business problems. M° Biagio's problems have reached us only in the transcription of M° Benedetto, who from folio 388^r to 408^r of **L.IV.21** quotes 114 of them. Furthermore M° Benedetto recommends presenting them in the same order in which they were written by the original author. Twenty-six of these problems are of a business type: eight concern travel, eight are calculations of interest, six require the determination of the price and the quantity of goods, and finally only one problem is devoted to each of the following subjects: alligation, calculation of wages, exchange of money.

The first business problem proposed by M° Biagio concerns the purchase of a certain number of *canne* of cloth; that of exchange concerns changing *fiorino* into *aghontani* and *bolognini*.

The interest of Biagio's solutions from an algebraic point of view lies in the suitable choice of the unknown and in the algebraic calculations necessary to obtain the equation in one of the canonical forms.[25]

---

[25] Further details on Biagio's problems can be found in R. Franci and L. Toti Rigatelli, 'Maestro Benedetto', *op. cit.*, pp. 305-307. All the problems are published in M° Biagio, *E' chasi, op. cit.*.

The other collections of problems are contained in manuscripts **Ricc. 2263, L.IX.28, Pal. 573, Ott.Lat.3307, Fond.prin.II.III.152**. In the Riccardiano manuscript, the first fourteen rules of the twenty-four listed are illustrated by one problem each. Afterwards, a collection of thirty-nine problems follows; twenty of them are of a business type. Some theoretical problems have in this case the manifest aim of teaching calculation with algebraic fractions. The first of these problems tells:

A man divides 100 by a quantity and afterwards he divides 100 by 5 more than the first time, and adding together these two results gets 20. We ask by what he divided at first?[26]

The other problems are similar, and some of them even consider three fractions.

In manuscript **L.IX.28**, scattered about the text, we can find forty-five problems solved by algebra, of which twenty are of a business type.[27]

Manuscripts **L.IV.21, Pal. 573, Ott.Lat.3307** contain, with some differences, passages from the most interesting collection of algebraic problems written in the fourteenth century.[28] In all we can list fifty-five different problems attributed by the authors of the above-mentioned manuscripts to the Florentine M° Antonio de´ Mazzinghi. The difficulty of these problems and the polish of their solutions can seldom be found even in the problems of the next century. It is sufficient to note that among the elements that characterize them are the use, probably for the first time, of two unknowns, and the solution of symmetric systems of the sixteenth degree in four unknowns. These systems are naturally reduced, by non-trivial devices, to quadratic equations or to binomial equations as high as the fifth degree.[29]

Also, each of the twenty-five rules of the manuscript **Fond.prin.II.III.152** is illustrated by one or two problems, and these are followed by forty-two others, whose solution is sometimes very involved.

It is precisely by reading the above-mentioned collections of problems that one can realize how in Italy little more than one century after the *Liber abaci*, algebra had already developed sufficiently to permit the authors a free and easy use of the calculation of rationalization processes, radicals, and algebraic fractions, while completely omitting, among other things, the geometric references so dear to Leonardo Pisano.

---

[26] '*Uno parte 1000 in una quantità e poi parte 100 in più 5 che prima e giunti questi due avenimenti insieme facie 20.*' Ricc. 2263, c.39^v.

[27] Gilio's problems are published in M° Gilio, *Questioni d'algebra, op. cit.*.

[28] The collection of Antonio's problems was entitled *Fioretti*. The selection from *Fioretti* made by M° Benedetto is published in Antonio de Mazzinghi, *Trattato di Fioretti, op. cit.*.

[29] For a broader analysis of Antonio's problems see R. Franci, 'Antonio de' Mazzinghi', *op. cit.*.

On an algorithm for the approximation of surds from a Provençal treatise

JACQUES SESIANO

1. Extant Provençal mathematical literature

The only extant testimony of mathematical activity in the Provençal language, besides marginal annotations in a few manuscripts, consists of three treatises: two printed books and one manuscript.

1.1 The printed books

The earlier of the two books was written by Frances Pellos—*citadin de Nisa*, as we read in the colophon—and was printed in Turin in 1492. The second book, also written by an inhabitant of Nice, Johan Frances Fulconis, was printed in Lyon in 1562. As was the case for most mathematical treatises of that time, both works were written for people primarily interested in commercial arithmetic. But, though not principally theoretical, these two books are not mere lists of rules illustrated by solved examples, as are some contemporary arithmetics: readers were considered capable of understanding the theoretical basis of the rules which was explained and discussed accordingly. In short, both these books were intended for an educated public and compare favourably with the majority of contemporary mathematical treatises.

1.2 The manuscript

The third extant Provençal text on mathematics is a manuscript copy of an original written by an unknown author around 1430 in the Languedocian city of Pamiers. It is now kept in the Bibliothèque Nationale (**Ms.fr.n**elles **acq.4140**), and it has been analysed only quite recently.[1] Like the two books, it is practical in aim; but here, even more than in the other two works, theoretical explanations are abundant, and the whole treatise is remarkably well presented. In addition, two important innovations grant it a special place in the history of mathematics.

---

[1] See our summary study of its contents. Its glossary may be useful for quotations found below.

1. It is the first known mathematical text in which a negative number is accepted as the solution to a problem. Two centuries earlier, Leonardo of Pisa had solved problems with negative results, but he had either declared these problems insoluble or, by changing the meaning of the concrete magnitude concerned, had transformed the negative result into a positive one.[2]
2. It explains an algorithm for the approximation of surds to any desired degree of accuracy, which, if not ideal for practical calculation, is quite interesting from a theoretical point of view, for it reveals the progressive approach towards a surd by a sequence of rational fractions, and it shows how any number of rationals can be inserted between any two given fractions. We shall examine this algorithm in detail below.[3]

1.3 Relationship of these three treatises

Curiously, we have no convincing evidence of a direct link between these three Provençal treatises.

We have some information about Fulconis' sources through a few references he makes to other works. The more or less contemporary treatises he mentions are of French or Italian origin, most notably the voluminous works of Pacioli (his *Summa*) and of de la Roche. There is no allusion to his fellow-townsman Pellos. As it happens, many problems in the two treatises are very similar, if not identical (this is true for the Provençal manuscript also); but since the same problems occur in contemporary French or Italian works too, occasional similarity in content or even in wording is no proof of filiation.

Pellos's case is even more problematic. He mentions no authors and no treatises. He has many problems not found in the treatise of Pamiers; on the other hand, some parts of his work (excepting dialectical differences) closely resemble some places in the manuscript. Of itself, this is not necessarily significant, as explained above, although the presence in Pellos's book of the problem with a negative solution in practically the same words tends to suggest at least indirect influence.

2. Approximation of surds in the treatise of Pamiers

After presenting the two sections on the extraction of the roots of square and cubic numbers, the author of this treatise explains the theory of the extraction of imperfect roots. At the beginning of this new section is given a general statement to the effect that any number which is neither square nor cubic possesses a root which is imperfect. In the subsequent theory, however,

---

[2] See our study on the occurrence of problems with negative solutions in medieval mathematics.

[3] Because of space limitations, we could not examine all its implications in our short study of the treatise of Pamiers.

the author will consider only imperfect roots of integers, the case of fractions being mentioned briefly at the end. In what follows, we shall present the author's arguments, quoting some important passages in the footnotes.

2.1 Theoretical foundations (fol. 66^r–69^v)

Any non-square/non-cubic integer is comprised between two integral squares/cubes, the roots of which are consecutive integers. In modern notation, if N is the non-square/non-cubic integer, we have

$$a^{2,3} < N < (a+1)^{2,3}$$

with a integral.

The (extracted) root of such an intermediate integer (*nombre entier mejancier*) N is referred to as an 'imperfect root' (*raditz emperfieyta*) because, when squared/cubed, it does not produce N; by definition, the number produced is supposed to differ from N by less than one.

The basic characteristics of imperfect roots are established in two theorems and certain conclusions drawn from them.

Theorem I The perfect root of any intermediate integral number N, $a^{2,3} < N < (a+1)^{2,3}$, can never be found.[4]

The demonstration rests on the following two lemmas:

L_1 Since $a^{2,3} < N < (a+1)^{2,3}$, the (imperfect) root of N must be the sum of a and some proper fraction.

L_2 No power of a (reduced) fraction can produce an integer.[5]

Proof(s) of Theorem I[6] Any imperfect (square or cubic) root of an intermediate integer N must consist of an integer and a proper fraction (by L_1); but squaring/cubing it cannot yield the integer N (by L_2). Whence the conclusion.

Theorem II As good an approximation as an imperfect root may be, one can always find a closer and better one.[7]

Lemmas:

L'_1 Unity is like a continuous quantity, which is infinitely subdivisible, and it does not possess a smallest part, since the denominator can be increased without end; (aliquot) fractions are endless by way of diminution as integers are by way of augmentation.[8]

---

[4] '*Jamais de dengun nombre entier contengut entre 2 nombres (entiers) prochans que an raditz perfieyta hom non atrobaran la raditz perfieyta.*' Theorem I and its proof were known in antiquity: cf. Eutocius' Commentary upon Archimedes' *Dimensio circuli* (Heiberg), III, p. 232, 9–13.

[5] This lemma is unclearly expressed in the extant text, but its signification can be inferred from its use later on.

[6] The text gives two proofs, but they really differ only in the wording.

[7] '*Per bona raditz que hom aja, totztemps la podes aver plus propdana e milhor.*'

[8] '*1 es como quantitat continua, que se pot metre en infinidas partidas ni se poden assignar las partidas en las quals derrierament 1 se pot metre, la qual causa se manifesta per lo denominador, lo qual per grant que ell sie, hom ne pot donar 1 major. Et aixi com lo nombre entier non ha fi en augmentatio, lo nombre rot non ha fi en diminutio*'. Note the correspondence $\{n\} - \{\frac{1}{n}\}$.

$$\frac{m_i a_1 + n_i a_2}{m_i b_1 + n_i b_2} \quad m_i, n_i \text{ natural.}$$

The first application of the rule of intermediate terms explained by Chuquet is to the solution of a quadratic equation by successive approximations. His example is well-chosen indeed, as only a few steps are needed to reach the exact value.[20] The second and principal application of the rule, to the extraction of imperfect roots, appears further on in the *Triparty*, after the explanation of the procedure for the extraction of exact roots.

The first part of Chuquet's algorithm for the extraction of imperfect roots is the same as in the Provençal treatise: by means of the two natural sequences of fractions, an upper and a lower limit are established. However, whereas the Provençal author introduces a multiplier h and thus obtains $h - 1$ intermediate fractions, Chuquet repeatedly applies his rule of intermediate terms, thus reducing each time the length of his interval from one side, until the desired degree of approximation is attained.

As an example, Chuquet computes $\sqrt{6}$. He also gives a list of impressively close approximations to the square roots of the (non-square) integers from 2 to 14, which '*par la rigle des moyens ont este trouvees*'. Having done so, he experienced the problems connected with the application of an approximation procedure; this makes his declared reluctance more understandable.

Remark In point of fact, the rule of intermediate terms was known in ancient times: Pappus has a proof of it in his *Collectio* (VII, prop. 8). Also, it was used in Islamic times in connection with the approximation of a root (*infra*, §6.1, *in fine*). However, we have no reason to doubt that Chuquet's claim to originality was made in good faith.

Turning to the case of (non-square) fractions, Chuquet suggests that one transform the given fraction

$$\frac{k}{l} \text{ to } \frac{k^2}{kl} \text{ or } \frac{kl}{l^2}$$

which makes it unnecessary to extract an imperfect root twice (as was explained in the Pamiers treatise).[21] Here too, Chuquet's great concern with practicability is patent.

The extraction of imperfect cubic roots is very briefly mentioned by

---

[20] The equation is $x^2 + x = 39\frac{13}{81}$, whence $5 < x < 6$. Then, using the *progression naturelle de augmentacion* and computing, he finds successively that $5\frac{1}{2} < x$, $5\frac{2}{3} < x$, $5\frac{3}{4} < x$, but $5\frac{4}{5} > x$. Thus, by the rule of intermediate terms, the value $5\frac{7}{9}$, which is the exact result (p. 654).

[21] '*Pour eviter la peine et l'ennuy que l'on peult avoir a sercher les racines de telz nombres (routz) qui ne sont quarrez ne d'ung (coste) ne d'aultre, l'on peult faire quarre l'ung ou l'aultre, lequel que l'on veult, en ceste maniere comme de $\frac{5}{7}$ (...) est plus facile de besongner sus $\frac{25}{35}$ ou sus $\frac{35}{49}$ que sus $\frac{5}{7}$, pour que aux deux premiers ne fault sercher si non une racine inparfaicte, et a $\frac{5}{7}$ en convient sercher deux. Et par ceste maniere peult on esquarrir tous aultres nombres routz*' (pp. 698–699). Note that such a rule was employed earlier, by Oriental as well as Occidental mathematicians (see Hunrath, pp. 26–28, 31; Tropfke, pp. 272, 280).

Chuquet, after his explanation of the procedure for exact roots. The better part of this short section is devoted to Chuquet's adamant statement of his distaste for a procedure so wasteful of time and energy.[22]

3.2 Transmission of Chuquet's algorithm

Chuquet's procedure for approximating surds was reproduced in its entirety (as were many other chapters of the *Triparty*) in the two editions of Estienne de la Roche's *Arismet(h)ique* (1520^1, 1538^2). It was thus available to sixteenth-century mathematicians, few of whom, however, seem to have paid it any marked attention, probably because of the tedious computations involved.

Juan de Ortega is the author of an arithmetical treatise entitled *Conpusicion de la arte de la arismetica y juntamente de geometria*, printed in 1512, which was reissued several times and was translated into French. From the edition of 1534 on, entitled *Tratado subtilissimo de arismetica y de geometria*, we find a group of approximations of surds, and Tannery has observed that they are obtainable by Chuquet's rule.[23] Note that these approximations occur only in editions later than the first issue of de la Roche's work.

Johannes Buteo [Jean Borrel] definitely knew Chuquet's procedure, since he explained it in his *Logistica* (printed in 1559).[24] By way of illustration, he computes approximate values of $\sqrt{13}$ and of a *numerus cum particula*, namely $\sqrt{13\frac{1}{2}}$. His source is no doubt de la Roche's work, which he mentions in the introduction to his *Logistica*.[25]

The well-read Juan Perez de Moya was certainly acquainted with Ortega's treatise (one of his works was printed together with a reissue of the latter's *Tratado subtilissimo* in 1563) as well as, presumably, contemporary French mathematical literature. Whichever his source, he explained Chuquet's procedure in his *Arithmetica practica y speculativa*.[26]

---

[22] '*Les racines cubiques imparfaictes, c'est assavoir des nombres qui ne sont pas vrays cubicz, se pevent sercher par la forme et maniere que l'on quiert les racines quarrees imparfaictes, combien que ce n'est que temps perdu et labeur sans utilite ne aulcune neccessite; car telles racines, puis qu'elles ne se pevent abreuier ne extraire, on les doit laisser ainsi qu'elles sont et les noter ainsi comme a este dit*' (p. 703).

[23] On the editions of Ortega's book and on these approximations, see Perott's article. Tannery's article also rectifies a printing mistake in one of the numerical approximations.

[24] See the above-mentioned study of Tannery or the *Logistica*, pp. 76–78. For cubic roots, Buteo gives only an approximation formula (see §6.2.2,**II**).

[25] '*(. . .) Lucas Italus* (Pacioli), *qui longe omnium optime simul et copiosissime scripsit idiomate suo. Quem proxime doctrina sequitur Stephanus a Rupe Lugdunensis, in opere suo lingua nostra Gallica composito*' (*Logistica*, pp. 6–7).

[26] See Wertheim's study, p. 150. Perez de Moya's book, first printed in Salamanca in 1562, was reprinted some twenty-five times, the last reissue, of a version *nuevamente corregida y añadidas muchas cosas*, having been printed in Madrid in 1798 (cf. Palau's *Manual*, XIII, pp. 94–95). Wertheim used the 1590 edition.

Finally, we are told by Adriaan Adriaansz (Metius) that his father Adriaan Anthonisz inferred from the limits

$$\frac{333}{106} < \pi < \frac{377}{120}$$

that a closer value is $\pi \cong \frac{355}{113}$.[27] He may well have learned the rule of intermediate terms from the *Logistica* of Buteo, whose name he mentions in a work published in 1589.[28]

4. A discussion of the nature of surds

We have seen that, while a related form of the Provençal algorithm had survived thanks to Chuquet (and de la Roche), the introductory, theoretical considerations found on fol. 66^r–69^v of the Provençal manuscript had not been transmitted and had thus fallen into oblivion. But the question of the nature of surds is a central one, bound to be raised again. Indeed, it was taken up by the prominent sixteenth-century German mathematician Michael Stifel (ca. 1487–1567) in his most important work, first printed in 1544, the *Arithmetica integra*.[29]

At the beginning of the second Book of this treatise, Stifel opens his study of Euclidean irrationals with a chapter entitled *De essentia numerorum irrationalium*. He first reminds us that there is disagreement as to whether irrationals (what he actually has in mind is: irrational roots of integers) are true numbers or not ('*disputatur de numeris irrationalibus, an veri sint numeri, an ficti*'). To prove that they are, some mathematicians argue that, in geometrical figures, they may represent quantities not expressible by rational numbers. On the other hand, notes Stifel, these irrationals—like the *infinitus numerus*, which is not a number—cannot be expressed with precision and do not bear a definite ratio to true numbers.[30]

To prove that these irrationals are neither integral nor rational, Stifel proceeds as had the Provençal author, pointing out that they are not integral

---

[27] See Mansion's note on this occurrence or his source, Bierens de Haan's study on Metius' father, pp. 3–5.

[28] See Bierens de Haan's study, pp. 9, 11.

[29] On Stifel and his *Arithmetica integra*, see J.E. Hofmann's study. [Addition to note 61, p. 21: Paul Eber (1511–1569) was a student of Stifel's friends Luther and Melanchthon and was at the University of Wittenberg at the same time as Stifel (1541)].

[30] '*Ubi eos* [sc. numeros irrationales] *tentaverimus numerationi subiicere, atque numeris rationalibus proportionari, invenimus eos fugere perpetuo, ita ut nullus eorum in se ipso praecise apprehendi possit (. . .). Non autem potest dici numerus verus, qui talis est ut praecisione careat, et ad numeros veros nullam cognitam habeat proportionem. Sicut igitur infinitus numerus non est numerus, sic irrationalis numerus non est verus numerus, quod lateat sub quadam infinitatis nebula sitque non minus incerta proportio numeri irrationalis ad rationalem numerum quam infiniti ad finitum*' (fol. 103^r).

40   Italian and Provençal mathematics

since they lie between consecutive integrals, and that they are not rational because no power of a fraction yields an integer. Stifel next shows, moreover, that between any two consecutive integers, say 2 and 3, there are infinitely many rational as well as irrational numbers. He does this by constructing these two sets:[31]

- for the rationals, he writes:

  $2\frac{1}{2}, 2\frac{1}{3}, 2\frac{2}{3}, 2\frac{1}{4}, 2\frac{3}{4}, 2\frac{1}{5}, 2\frac{2}{5}, 2\frac{3}{5}, 2\frac{4}{5}, 2\frac{1}{6}, 2\frac{5}{6}, 2\frac{1}{7}, 2\frac{2}{7}, 2\frac{3}{7},$ '*et sic deinceps in infinitum*'
  [i.e. $\{2 + \frac{i}{k}\}$, with $1 \leq i \leq k - 1$, where k and i natural, and leaving out the repetitions];

- for the irrationals (surds), he has:

  $\sqrt{5}, \sqrt{6}, \sqrt{7}, \sqrt{8}, \sqrt[3]{9}, \sqrt[3]{10}, (\ldots), \sqrt[3]{26}, \sqrt[4]{17}, (\ldots), \sqrt[4]{26},$ '*et sic deinceps in infinitum*'
  [i.e. $\{\sqrt[n]{r_{n,i}}\}$, with $r_{n,i}$ integer such that $2^n < r_{n,i} < 3^n$, thus $i = 1, \ldots, 3^n - 2^n - 1$].

Stifel remarks that, although the two sets are infinite, they possess no common element.[32]

5. An argumentation against the comparability of infinites in medieval philosophy

5.1 Historical survey[33]

In raising the question of the possibility of the infinite, Aristotle was led to distinguish between two types of infinites: the *potential* infinite, which involves an endless succession, the ultimate 'part' of which can never be reached; the *actual* infinite, which is a completed infinite. While he denied the possibility of an actual infinite, mainly on the basis of physical arguments such as its being neither composite nor simple, Aristotle admitted the potential infinite in three cases: the infinitely divisible *continuum* (time or space), the infinity of *time*, the infinity of natural *numbers*—which, like time, is a potential infinite by way of addition.

Theological considerations led medieval philosophers to reconsider Aristotle's statements about infinity. Given the dogma of God's

---

[31] His construction gives all rationals between 2 and 3, and all irrational roots of integers between 2 and 3. His disposition allows us to enumerate them.

[32] '*Item licet infiniti numeri fracti cadant inter quoslibet duos numeros immediatos, quemadmodum etiam infiniti numeri irrationales cadunt inter quoslibet duos numeros integros immediatos. Ex ordinibus tamen utrorumque facile est videre, ut nullus eorum ex suo ordine in alterum possit transmigrare. Nihil igitur est, si cogites numerum aliquem irrationalem posse coincidere cum aliquo numero fracto propter infinitatem fractorum*' (fol. 103^v–104^r).

[33] The subject is discussed in various places of Duhem 1906–1913, in Maier 1949, pp. 155–215 and 1964, pp. 41–85, and in Murdoch 1982.

Assumption 2 (A_2): If, when two things are superposed, or applied to one another *per imaginationem*, the one does not exceed or is not exceeded by the other, neither one is larger or smaller.

Similarly, a multitude is neither larger nor smaller than another when 'they are such that to any unit of the one corresponds a unit in the other'.[45]

Assumption 3 (A_3): There is between infinite and finite no finite proportion.

Otherwise, taking a finite a finite number of times would produce an infinite.[46]

Theorem I (T_1): No infinite is larger or smaller than another infinite.[47]

In order to show that neither of two seemingly different infinites of the same kind is larger, Albert gives the following three arguments:

α) Suppose A to be a body infinite from every side, and B to be a body with a cross-section of one square foot but infinitely long in one direction. Considering B, assume the following to be done in the proportional parts of one hour. Take from B a cubic foot, then give it a spherical shape; take from B a second cubic foot, put it together with the previous one, and give the whole a spherical shape; take next a third cubic foot, mould this and the sphere so as to form a new sphere; and so on. At the end of the hour, we shall have a body which is infinite from every side, as the result of an infinite number of equal additions. This body can thus be applied to A and, both being infinite from every side, B neither exceeds A nor is exceeded by it; hence B is neither larger nor smaller than A (by A_2). Now, B as a whole has not been changed quantitatively during the above operation (A_1). Hence, in its original form, B must have been neither smaller nor larger than A.[48]

---

[45] '*Quecumque sibi invicem superposita vel per imaginationem applicata sunt si unum non excedit aliud nec exceditur, unum eorum nec est maius nec est minus alio. Et similiter est de multitudinibus que sic se habent quod cuilibet unitati in una correspondet unitas in alia earum, una non est maior altera neque minor.*'

[46] '*Infiniti ad finitum nulla est proportio finita, sicut dupla vel tripla etc.; quia tunc sequeretur quod finitum esset medietas infiniti vel pars aliquota ita quod finities sumpta faceret infinitum.*'

[47] '*Nullum infinitum altero infinito est maius aut minus, scilicet corpus corpore, superficies superficie, multitudo multitudine, tempus tempore, virtus virtute.*'

[48] '*Per imaginationem, si est possibile, sit a unum corpus infinitum undique et b sit unum aliud infinitum pedaliter latum et pedaliter profundum et in infinitum longum solum versus occidentem. Tunc per imaginationem dimisso a sumatur primum pedale de b, et fiat sphericum, deinde secundum addatur primo et fiat totum sphericum, deinde tertium, deinde quartum, et sic ultra; et non fiat rarefactio nec aliunde additio. Et continuetur hoc secundum partes proportionales hore, quod adversarius non reputaret impossibile. Tunc in fine hore totum hoc erit infinitum undique, quia infinite sunt facte additiones equales. Ponatur ergo per imaginationem quod applicetur ipsi a, ac si duo corpora essent simul. Tunc non excedit a nec exceditur, quia quodlibet undique est extensum. Ergo per secundam suppositionem non erit maius neque minus. Sed ipsi b nihil est additum, nec ipsum est rarefactum. Ergo per primam suppositionem non est augmentatum. Ergo in principio quando erat infinitum solum secundum longum, et ab una parte non erat maius quam a neque minus. Sed tamen si aliquod infinitum esset vel posset esse maius alio, a deberet esse maius (quam) b.*' Cf. also Duhem 1906, pp. 343-344.

46 *Italian and Provençal mathematics*

After having remarked that assumption A_2 can be similarly used for any pair of infinites, Albert notes that, by the same representation, one can show that from B can be created an infinite number of infinite globes: take every other (evenly numbered) cubic foot and form an infinite globe as before; then join together the remaining pieces of B. Since they are infinite in number, the same operation can be repeated as many times as desired.[49]

β) Let I, I' be two infinites, and ΔI their difference. Consider the ratio ΔI bears to (either one of) the infinites.

First, let us suppose ΔI to be finite. As the said ratio cannot be finite, according to A_3, it must be infinitely small, and one of the infinites is not properly larger than the other, according to D_2.

Suppose next ΔI to be infinite. The ratio cannot be finite here either, 'for then one infinite would be (e.g.) half another infinite, the opposite of which is shown (to hold) in the third (book) of the Physics [III.5, 204a]'. Nor may the ratio be infinite, according to D_2.[50]

γ) Suppose a measure of one foot be divided into proportional parts. On the first part, let there be 'something white', on the second something black, on the third something white again, and so on alternately. Let us remove the first black thing and put in its place the second white one, and let us go on and remove all the blacks replacing them successively with whites. Clearly there will be something white on every part, and thus the multitude of the whites is neither larger nor smaller than the multitude of the parts; but the same held previously for both the whites and the blacks together, so that the multitude of the whites and blacks is neither larger nor smaller than that of the whites alone.[51]

---

[49] '*Iuxta eandem imaginationem sequitur quod ex b potest fieri infinitum undique, et tamen b adhuc non esset diminutum. Ut si capiatur secundum pedale, et deinde quartum, et deinde sextum, et deinde octavum, et sic ulterius secundum partes alternatas, et ponantur in globo sicut prius, et alia pedalia remanentia coniungantur, tunc patet propositum. Nec solum hoc sequitur, sed etiam quod ex ipso b non diminuto potest fieri adhuc unum aliud infinitum per eandem viam, immo infinities infinita.*' After the creation of the nth infinite, there remains the infinite set of those *pedalia* which originally bore the numbers $1 + k.2^n$, $k = 0,1,2,\ldots$

[50] '*Si unum infinitum excederet aliud, hoc esset aut excessu finito aut infinito. Non finito, quia tunc esset proportio in infinitum parva—quod patet, quia si esset certa proportio sicut dupla, tunc finitum esset medietas infiniti, vel pars aliquota, quod est contra tertiam suppositionem; et si est in infinitum modica, tunc unum non est maius alio proprie, sicut patet ex secunda distinctione. Nec potest dici quod unum infinitum excedat aliud excessu infinito, quia aut hoc esset finite, scilicet in certa proportione sicut dupla vel tripla etc., aut infinite ultra omnem proportionem finitam. Non potest dici primum, nam tunc unum infinitum esset medietas alterius infiniti, cuius oppositum ostenditur tertio Physicorum; nam (. . .) non posset assignari ubi esset medium, sicut patet de tempore eterno. Nec potest dici secundum, nam tunc non esset proprie maius, sicut dicebatur in secunda distinctione, sicut nec corpus est proprie maius superficie nec superficies linea.*'

[51] '*Arguitur de multitudine. Nam sit unum pedale per imaginationem divisum per partes proportionales. Tunc super primam partem sit aliquod album, super secundam sit aliquod*

On an algorithm for the approximation of surds from a Provençal treatise

The same reasoning can be applied for an initial disposition of the whites on (say) every thousandth part, counted from the first one on.[52]

Theorem II is based on the three following assumptions:

Assumption 1 (A'_1): A part of some thing is not equal to the (homogeneous) whole from which it is taken.[53]

Assumption 2 (A'_2): If a first thing exceeds some other thing and is not exceeded by it by any part, these two are not equal.[54]

Assumption 3 (A'_3): If from equals/to equals are removed/added (un)equals, the results are (un)equal, whether the removed/added parts bear a ratio to the results or not.[55]

Theorem II (T_2): Pairs of unequal infinites do exist.[56]

Arguments (compare with α, β in T_1):

α') Let A be a body infinite from every side. Imagine in it a part B consisting of a beam with a cross-section of one square foot and infinitely long in one direction ('*versus occidentem*'). Then the two infinites A and B are not equal (by A'_1).[57]

β') Let A and B be two infinite beams as above, the one not exceeding the other (in the finite). Add to A a body C of four cubic feet and to B a body D of one cubic foot. Then, either A and B are not equal, or they are equal but A + C and B + D must be unequal (by A'_3). Either way, the proposition is verified.

---

nigrum, et super tertiam iterum sit aliquod album, et super quartam iterum aliquod nigrum, et sic alternatim de aliis partibus proportionalibus. Tunc auferatur primum nigrum, et transferatur secundum album super secundam partem; deinde auferatur tertium nigrum, transferendo sicut prius. Et consequenter amoveantur omnia nigra. Istum casum concederent adversarii. Tunc clarum est quod super quamlibet partem erit aliquod album. Ergo per primam [secundam] suppositionem multitudo alborum non est maior nec minor multitudine partium proportionalium. Et per idem patet ex principio casus quod tota multitudo alborum et nigrorum non est maior nec minor multitudine partium proportionalium. Ergo sequitur quod alba et nigra simul sumpta non sunt plura nec pauciora quam alba solum.'

[52] *'Eodem modo si super primam ponatur unum album, deinde transcendendo ponatur secundum album super millesimam, deinde super millesimam ab illa, et sic consequenter, probabitur ut prius quod partes proportionales non sunt plures quam illa alba.'*

[53] *'Portio alicuius non est equalis illi cuius est portio, et cum quo est eiusdem rationis. Hoc videtur per se notum; quia, si negaretur, eodem modo posset dici quod pars et illud cuius est pars essent idem vel equalia.'*

[54] *Si unum superexcedit et transcendit aliud et non ab aliqua parte exceditur ab illo, illa non sunt equalia.'*

[55] *'Si ab equalibus equalia demas, que remanent erunt equalia, sive dempta habeant proportionem ad residua sive non. Sicut si a duobus angulis rectis removeantur duo anguli contingentie equales, restant equalia. ¶Et iterum si equalibus iungantur inequalia, tota fiunt inequalia, sive habeant proportionem sive non, sicut prius. Et non est ratio quare ista sint magis vera de finitis quam de infinitis, si essent.'*

[56] *'Aliqua duo infinita non sunt equalia.'*

[57] *'Sit a unum corpus infinitum undique (. . .). Tunc in eo potest signari portio que est pedaliter lata et profunda et in infinitum longa solum versus occidentem, que sit b. Tunc arguitur sic: a et b non sunt equalia, ergo aliqua duo infinita non sunt equalia; antecedens patet, quia b est portio a, ergo per primam suppositionem b et a non sunt equalia.'*

48 *Italian and Provençal mathematics*

One comes to the same conclusion by imagining that A is moved *ad orientem* until its extremity (in the finite) goes further than that of B.[58]

The same reasoning can be used for infinite bodies of other shapes (*non tamen undique*, specifies Albert). It is also applicable to infinite multitudes, or to qualities—which, when infinite, must produce an infinite effect.[59]

Theorem III: No infinites are comparable one to another.

For, no infinite is larger or smaller than another (T_1), so that they are either all equal or all not comparable. Since not all are equal (T_2), they must all be not comparable.[60]

The Question ends with various inferences from the previous theorems, in particular a set of corollaries on the impossibility of increasing, diminishing, or parcelling out an actual infinite, if such an infinite exist.[61]

6. Some other methods of extracting imperfect roots

We have seen that, after explaining his algorithm for the approximation of surds, the Provençal author declared it to be superior to the 'other rules' he knew of, for it gives both an upper and a lower limit, whereas the other procedures, he says, yield only one limit, a lower one. In order to determine which other rules he may have had in mind, we shall now consider the methods of approximating surds which were used in ancient and/or medieval times.

6.1. Square roots

6.1.1. Multiplying the radicand by a square

In order to calculate $\sqrt{N}$, N integral, choose an integer R large enough

[58] '*Sit a pedaliter latum et profundum et in infinitum longum versus occidentem solum, et b simili modo, et superponantur ita quod unum non transcendat alterum. Tunc a et b videntur esse equalia, si aliqua infinita sunt equalia. Deinde iungatur ipsi a corpus quatuor pedum, et sit c; et ipsi b iungatur d corpus unius pedis. Tunc arguitur. Si a et b non sunt equalia, habetur intentum. Si sint equalia, ergo per tertiam suppositionem adiunctis c et d inequalibus tota fiunt inequalia, ergo duo infinita, scilicet a, c ex una parte et b, d ex alia, sunt non equalia. Quod est propositum. ¶Confirmatur. Sint a et b corpora ut prius, et trahatur a ad orientem usque quo a transeat extremitatem b. Tunc a superpositum b transcendit ipsum, et non e converso. Ergo a et b non sunt equalia.*'

[59] '*Consimiles rationes adduci possunt de multitudinibus infinitis. Consimili modo potest argui de virtutibus, quia non dicuntur infinite* [cathegorice, add ms.] *nisi quia possunt producere infinitum effectum, et ideo ex non equalitate effectuum arguetur non equalitas causarum vel virtutum. Concludatur ergo quod in quolibet genere in quo possunt imaginari infinita aliquid alteri est non equale.*'

[60] '*Nulla infinita sunt adinvicem comparabilia. ¶Probatur. Nullum infinitum est maius alio nec minus per primam conclusionem. Ergo vel omnia sunt equalia, vel non sunt adinvicem comparabilia. Sed non omnia sunt equalia per secundam conclusionem. Ergo omnia non sunt adinvicem comparabilia. Et hoc dicit conclusio.*'

[61] '*Infinitum non potest augeri nec addendo finitum nec addendo infinitum (. . .) nec potest minui (. . .) nec potest duplari nec triplari (. . .) nec etiam subduplari vel subtriplari (. . .); sequitur quod infiniti non est pars proprie dicta, nec aliquota (. . .) nec etiam non aliquota (. . .).*'

(generally of the form 10^n), and extract the integral root of $N.R^2$ by the usual method. Then

$$\sqrt{N} \cong \frac{1}{R}\sqrt{NR^2}$$

The size of R will be taken according to the desired degree of accuracy.

This method appears in Indian mathematics, in Islamic mathematics from the tenth century on, and then in Europe, as early as the twelfth and thirteenth centuries (J. Hispalensis, Leonardo of Pisa).[62]

6.1.2. Approximation formulae

Consider N integral, $a^2 < N < (a+1)^2$; thus

$$\sqrt{N} = a + x \qquad 0 < x < 1$$
$$N = a^2 + r \qquad 0 < r = 2ax + x^2 < 2a + 1$$

Since $2ax < r < 2ax + x$, we have

$$\frac{r}{2a+1} < x < \frac{r}{2a}, \text{ whence}$$

$$a + \frac{r}{2a+1} < \sqrt{a^2 + r} < a + \frac{r}{2a}$$

6.1.2.1. Sequence of approximations by excess

The approximation procedure

$$\sqrt{a^2 + r} \cong a + \frac{r}{2a} \equiv a_1$$

can be applied again for a_1, and so on repeatedly (this is Newton's method for $f(x) = x^2 - N$). The sequence

$$a_i \equiv a_{i-1} + \frac{r_{i-1}}{2a_{i-1}} = a_{i-1} - \frac{|r_{i-1}|}{2a_{i-1}},$$

with $r_{i-1} \equiv N - a^2_{i-1} = -\left(\frac{r_{i-2}}{2a_{i-2}}\right)^2$

approaches indefinitely close to $\sqrt{N}$ from the right side.

This type of approximation is first found in Babylonian mathematics; Heron explicitly stated that the procedure can be applied repeatedly (*Metrica* I.8).[63] It appears to have been in common use both among Oriental and

---

[62] See Treutlein, pp. 69–71; Hunrath, pp. 25–29, 43–44; Tropfke, pp. 272, 278–280.
[63] Heron uses the equivalent form

$$\sqrt{N} \cong a_i \equiv \frac{1}{2}\left(a_{i-1} + \frac{N}{a_{i-1}}\right),$$

in which $\sqrt{N}$, the geometric mean of a_{i-1} and $\frac{N}{a_{i-1}}$, is seen to be approximated by their arithmetic mean.

50   *Italian and Provençal mathematics*

Occidental mathematicians, the latter from the twelfth century on (J. Hispalensis; iteratively in Leonardo's *Liber abaci*). Cataldi (1552–1626) discovered the link between this sequence and the partial quotients resulting from the development of $\sqrt{a^2 + r}$ in a continuous fraction (the terms of the former are the 2^nth terms of the latter).[64]

6.1.2.2. Sequence of approximations by defect
The approximation procedure

$$\sqrt{a^2 + r} \cong a + \frac{r}{2a + 1} \equiv a_1'$$

can also be applied repeatedly. The results thus obtained, viz.

$$a_i' \equiv a_{i-1}' + \frac{r_{i-1}'}{2a_{i-1}' + 1},$$

with $r_{i-1}' \equiv N - a_{i-1}'^2 = \frac{r_{i-2}'}{2a_{i-2}' + 1}\left[1 - \frac{r_{i-2}'}{2a_{i-2}' + 1}\right] > 0,$

approach indefinitely close to $\sqrt{N}$ from the left side.

This type of approximation is believed to have been used in Greek times, but texts attesting its use do not occur before the tenth century in Islamic countries. Unlike the previous method, it is not found in Europe in the twelfth century or in Leonardo's works. Although it does occur in a manuscript written around 1300 (see below, §6.3.2), it appears to have been in common use only in the sixteenth century. Here again, the iterative process was studied in depth by Cataldi.[65]

Remarks
1. If we start from the upper integral limit $b = a + 1$, we have

$$b - \frac{s}{2b - 1} < \sqrt{b^2 - s} < b - \frac{s}{2b}$$

Some Greek results may have been obtained by the use of these limits. But we know with certainty that the approximation on the right side was used by Islamic mathematicians and by Leonardo.[66]

Again, since $0 < r, s < 2a + 1$ and the smaller r, s are, the better the approximation, one may prefer, in the case $N = a^2 + r$ with $r > a$, to apply the formula

$$\sqrt{N} \cong b - \frac{s}{2b}$$

---

[64] See Treutlein, pp. 68–69; Günther, pp. 29–30, 45–46, 57–58; Hunrath, pp. 27–30, 37–43; Tropfke, pp. 264–265, 267, 272, 277, 289; Saidan, p. 441; Maracchia, pp. 67 *seqq.*, 105.
[65] See Treutlein, pp. 67–69; Perott, p. 168; Günther, pp. 44–45; Hunrath, pp. 21, 26, 27, 35, 37; Carruccio, p. 127 or Maracchia, p. 75 *seqq.*; Tropfke, p. 277.
[66] See Hunrath, p. 32; Tropfke, p. 277.

where $b = a + 1$ and $s = 2a + 1 - r$. Inserting these values and computing, one obtains

$$\sqrt{N} \cong a + \frac{r+1}{2a+2},$$

a formula seen in Islamic sources.[67]

2. Since

$$a + \frac{r}{2a+1} < \sqrt{a^2 + r} < a + \frac{r}{2a},$$

we know by the rule of intermediate terms that a better approximation is

$$a + \frac{2r}{4a+1}.$$

This is how al-Uqlīdisī (tenth century) obtained this formula.[68] We have thus a simple application of the Pappus–Chuquet rule, midway between late Greek and late medieval times.

6.2. Cubic roots

6.2.1. Multiplying the radicand by a cube

In order to calculate $\sqrt[3]{N}$ we may, as in the case of square roots, consider

$$\frac{1}{R}\sqrt[3]{NR^3}$$

and extract in the usual way the cubic root of NR^3, keeping only its integral part. This is done by Islamic and sixteenth-century European mathematicians.[69] It is also found in medieval Latin manuscripts.[70]

6.2.2. Approximation formulae

Let N be an integer comprised between the integral cubes a^3 and $(a + 1)^3$. Then

$$\sqrt[3]{N} = a + x \quad 0 < x < 1$$

and $N = a^3 + r \quad 0 < r = 3a^2 x + 3ax^2 + x^3 < 3a^2 + 3a + 1$.

Since $3a^2 x < r < (3a^2 + 3a + 1)x$, we have

$$\frac{r}{3a^2 + 3a + 1} < x < \frac{r}{3a^2}.$$

---

[67] See Günther, p. 45; Hunrath, pp. 21, 27; Saidan, pp. 445, 448.
[68] See Saidan, pp. 164, 441.
[69] See Treutlein, p. 76; Saidan, p. 460; Tropfke, pp. 278–280.
[70] For instance at the very end of the *Brevis ars minutiarum* (see, e.g. **ms Montpellier 323**

52   Italian and Provençal mathematics

Thus, of the various approximations

$$a + \frac{r}{3a^2 + 3a + 1} < a + \frac{r}{3a^2 + 3a} < a + \frac{r}{3a^2 + 1} < a + \frac{r}{3a^2},$$
$$\text{(I)} \qquad\qquad \text{(II)} \qquad\qquad \text{(III)} \qquad\qquad \text{(IV)}$$

I is always by defect, whereas **IV** is always by excess (the two middle ones can lie on either side of the required value).

All these approximations are attested in medieval and Renaissance times:

I occurs in the writings of Islamic mathematicians, as well as Christian ones, beginning with Leonardo of Pisa.[71]

II has been used by Tartaglia and by Buteo.[72]

III was known to Islamic mathematicians.[73]

IV occurs in Arabic texts, and is employed iteratively by Cardan.[74] This is Newton's algorithm applied to $f(x) = x^3 - N$.

6.3. *Whether we have solved the question of the 'other methods' alluded to*

6.3.1. *Arguitur quod sic*

Among these methods of approximating square and cubic roots, two systematically give lower limits, and thus may have been those referred to by the Provençal author:

1 The method of multiplying the radicand by a square, respectively by a cube (since, in extracting the numerator's root, we dismiss the non-integral part, the resulting fraction will be in defect of the exact value).

2 The approximation

$$\sqrt[n]{a^n + r} \cong a + \frac{r}{(a+1)^n - a^n}$$

[saec. XIV], fol. 276vb, or **ms Sorbonne 1037** [saec. XV], fol. 214^v); the explanations (given here with minor changes) are: '*Integrorum quoque radicem (cubicam) cum ipsorum numerus non fuerit cubicus sic iuxta artem minutiarum poteris invenire. Numerum magnum quemcumque volueris propones, quem duces in se cubice, et ipsum cubum facies denominatorem; deinde in eundem cubum duces numerum integrorum, et productum facies numeratorem eiusdem minutie. Scias igitur hanc minutiam equalem esse integris datis. Ipsius ergo radicem modo predicto reperies. Sed iam invenisti radicem denominatoris. Tantum ergo queres radicem numeratoris, secundum propius aut minus propius, et inventam facies numeratorem radicis, et habebis propositum (. . .). Et scias quod quanto maior fuerit numerus propositus predictus, tanto propius radicem invenies; valde magnum (numerum) igitur proponas*'. The corresponding approximation for square roots is found just before (fol. 275va and fol. 213^v in the aforesaid manuscripts.)

[71] See Treutlein, p. 76; Hunrath, pp. 35, 38, 47; Tropfke, p. 278.

[72] See Hunrath, p. 38; Buteo, p. 83. Buteo introduces this approximation thus: '*Ad hoc autem falsas quidem regulas commenti sunt. Vera autem sic habet.*'

[73] See Tropfke, p. 278.

[74] See Treutlein, p. 76; Saidan, p. 463.

with $n = 2$ and $n = 3$.

6.3.2. Arguitur quod non

We cannot consider the problem solved in an entirely satisfactory way, for:
 1 while the Provençal author speaks of 'other rules', which implies at least two, we have been able to find at most two possible types;
 2 the formula

$$a_1 = a + \frac{r}{2a + 1}$$

is, unlike the corresponding one for the case $n = 3$, rarely attested in Europe before Renaissance times;[75]
 3 it is surprising to see the Provençal author apparently ignorant of the most common method for approximating square roots, that which produces only upper limits.[76]

BIBLIOGRAPHY

Albertus de Saxonia (1516). *Quaestiones et decisiones physicales insignium virorum Alberti de Saxonia (. . .) Thimonis (. . .) Buridani (. . .) in Aristotelis (. . .), recognitae summa accuratione et iudicio Magistri Georgii Lokert Scoti*. J. Bade & C. Resch, Paris.
Bierens de Haan, D. (1878) 'Bouwstoffen voor de geschiedenis der wis- en natuurkundige wetenschappen in de Nederlanden, XII: Adriaan Anthonisz', *Versl. en meded. aft. natuurk.*, 2de reeks, deel XII, pp. 1–35.
Busard, H. (1961) *Nicole Oresme: Quaestiones super geometriam Euclidis*. E.J. Brill, Leiden.

---

[75] An early occurrence, going back to the XIIIth or the early XIVth century, is found in the manuscript **Colmar, Bibl. mun. 414** (no. 365 in the *Catalogue* of 1969). Between Sacrobosco's *Algorismus* and his *De spera*, there is a set of complements on mean proportionals and roots, as well as recreational problems (fol. 8^v, 22–10^r). One reads on fol. 9^r, 12–25: '*Posito quod unum castrum distet ab alio per 4 leucas et aliud ab alio per 4 et distent ad angulum rectum, si vis scire quantum distent a se remotiora, facias de numeris propositis quadratos, et illos addas sibi adinvicem; deinde, si possis, invenias unum numerum qui ductus in se quadrate equivaleat illis quadratis, et ille numerus ostendet distantiam iam dictam. Si vero nullus numerus valeat inveniri qui precise reddat, in se ductus, illos duos quadratos, sume numerum qui reddat quadratum minorem proximum tantum, et vide quod* [sc. *quot*] *unitates desint in illo quadrato quod non comprehendit alios duos quadratos, et tot partes appones, de illis, dico, que provenirent per appositionem gnomonis.* ¶*Verbi gratia. Bis quater 4 sunt triginta duo. Nullus quadratus constituit ipsum. Accipe ergo minorem, scilicet 25, et constat quod 7 deficiunt. Si ipsi apponeremus gnomonem, resultarent per appositionem gnomonis undecim. Ergo 7 undecime defuerunt.*'

[76] The Heronian form of it (cf. note 63), which yields both an upper limit, a_i, and a lower limit, $\dfrac{N}{a_i}$, does not seem to have been in use in medieval times.

Buteo, J. (1559) *Logistica, quae & arithmetica vulgò dicitur.* G. Rouville [Rouillé], Lyon.

Carruccio, E. (1971) 'Cataldi', in *Dictionary of scientific biography* (ed. C. Gillispie), vol. 3, pp. 125-129. Charles Scribner's Sons, New York.

Duhem, P. (1906, 1909, 1913) *Etudes sur Léonard de Vinci* (3 volumes). A. Hermann et fils, Paris.

Fulconis, J. Fr. (1562) *Opera nova d'arismethica intitulada Cisterna Fulcronica.* Th. Bertheau, Lyon.

Galilei, G. (1655) *Discorsi e dimostrationi matematiche intorno à due nuove scienze* (2nd edn., in: *Opere, racc. da C. Manolessi,* vol. 2). HH. de Duciis [Eredi del Dozza], Bologna.

Gericke, H. (1977) 'Wie vergleicht man unendliche Mengen?' *Sudhoffs Archiv,* 61, pp. 54-65.

Günther, S. (1882) 'Die quadratischen Irrationalitäten der Alten und deren Entwickelungsmethoden.' *Abh. Gesch. Math.* 4, pp. 1-134.

Heiberg, J. (1910, 1913, 1915) *Archimedis Opera omnia cum commentariis Eutocii* (3 vols). B. Teubner, Leipzig.

Hofmann, J.E. (1968) *Michael Stifel 1487?-1567. Leben, Wirken und Bedeutung für die Mathematik seiner Zeit.* F. Steiner, Wiesbaden.

Hunrath, K. (1884) *Die Berechnung irrationaler Quadratwurzeln vor der Herrschaft der Decimalbrüche.* Lipsius & Tischer, Kiel.

Maier, A. (1949) *Die Vorläufer Galileis im 14. Jahrhundert.* Storia e Letteratura, Rome.

Maier, A. (1964) *Ausgehendes Mittelalter, I.* Storia e Letteratura, Rome.

Mansion, P. (1888) 'Note historique sur la règle de médiation.' *Bibliotheca math.,* 2.F, 2, p. 36.

Maracchia, S. (1979) *Da Cardano a Galois. Momenti di storia dell'algebra.* G. Feltrinelli, Milan.

Marre, A. (1880) 'Notice sur Nicolas Chuquet et son Triparty en la science des nombres.' *Bull. bibl. stor. sci. mat. e fis.* 13, pp. 555-659 & 693-814.

Meschkowski, H. (1964) *Wandlungen des mathematischen Denkens* (3rd edn.). F. Vieweg & Sohn, Braunschweig.

Murdoch, J. (1982) 'Infinity and continuity', in *The Cambridge history of later medieval philosophy* (eds. N. Kretzmann *et al.*) pp. 564-591. Cambridge University Press, Cambridge.

Pacioli, L. (1494) *Summa de arithmetica, geometria, proportioni e proportionalita.* P. de Paganini, Venice.

Palau y Dulcet, A. (1948-   ) *Manual del librero hispanoamericano* (2nd edn.). A. Palau, Barcelona.

Pellos, Fr. (1492) *Compendion de lo abaco.* N. de Benedetti & J. Suigo, Turin.

Perott, J. (1882) 'Sur une arithmétique espagnole du seizième siècle.' *Bull. bibl. stor. sci. mat. e fis.* 15, pp. 163-169.

de la Roche, E. (1520[1], 1538[2]) *L'arismet(h)ique*. Const. Fradin[1], G. et J. Huguetan[2], Lyon.

Saidan, A. (1978) *The Arithmetic of Al-Uqlīdisī*. D. Reidel, Dordrecht.

Sesiano, J. (1984) 'Une arithmétique médiévale en langue provençale.' *Centaurus* 27, pp. 26–75.

Sesiano, J. (1985) 'The appearance of negative solutions in mediaeval mathematics.' *Arch. hist. exact sci.* 32, pp. 105–150.

Stifel, M. (1544) *Arithmetica integra*. J. Petreius, Nürnberg.

Tannery, P. (1887) 'L'extraction des racines carrées d'après Nicolas Chuquet.' *Bibliotheca math.*, 2.F, 1, pp. 17–21.

Treutlein, P. (1877) 'Das Rechnen im 16. Jahrhundert.' *Abh. Gesch. Math.* 1, pp. 1–100.

Tropfke, J. (1980) *Geschichte der Elementarmathematik*, Bd. 1 (4th edn., eds. K. Vogel *et al.*). W. de Gruyter, Berlin.

Wertheim, G. (1898) 'Die Berechnung der irrationalen Quadratwurzeln und die Erfindung der Kettenbrüche.' *Abh. Gesch. Math.* 8, pp. 147–160.

PART II

Nicolas Chuquet and French mathematics

Nicolas Chuquet—an introduction

GRAHAM FLEGG

VARIOUS standard histories of mathematics make reference to the manuscript of Nicolas Chuquet, composed in 1484. However, with very few exceptions, these references are merely to that part of the manuscript known as the *Triparty*. In fact, the *Triparty*, discovered by Aristide Marre and published in the *Bullettino di bibliografia e di storia delle scienze matematiche e fisiche* in 1880, is only a part of a much larger manuscript which also includes a geometry and a commercial arithmetic as well as an appendix of problems, part of which was published in the *Bullettino* in 1881.[1] In 1979, M. L'Huillier published his authoritative transcription of the geometry, together with an introduction and notes on the original text.[2] Apart from Itard's article in the *Dictionary of scientific biography*,[3] there seems to be virtually no information in the English language about the content and importance of Chuquet's manuscript taken as a whole.

We can think of the manuscript as four works: the *Triparty en la science des nombres* (a treatise on arithmetic and algebra), a book of problems (including recreational problems) illustrating the application of the principles of the *Triparty*, a geometry entitled 'How the science of numbers may be applied to the measurements of geometry',[4] and a commercial arithmetic with the title 'How the science of numbers may be applied in matters of merchandise'.[5] The *Triparty* comprises the first 147 folios, the problems folios 148-210 (the last five being the recreational problems), the geometry folios 211-262, and the commercial arithmetic folios 264-321. The remaining pages consist of various tables and what Chuquet calls *les canons*. According to Chuquet himself, the *Triparty* at least was written in Lyon in 1484. At the end of the first work, he writes:

---

[1] Aristide Marre, 'Notice sur Nicolas Chuquet et son *Triparty en la science des nombres*', *Bullettino di bibliografia e di storia della scienze 13* (1880) pp. 555-592; 'Le *Triparty en la science des nombres* par Maistre Nicolas Chuquet, parisien', *ibid.*, pp. 593-658, 693-814; 'Appendice au *Triparty en la Science des nombres* de Nicolas Chuquet, parisien', *ibid.*, 14 (1881) pp. 413-460.
[2] Hervé L'Huillier, *Nicolas Chuquet, la Géométrie. Première géométrie algébrique en langue française (1484)*, Paris, Vrin, 1979.
[3] Jean Itard, 'Chuquet, Nicolas' in C. C. Gillespie (ed.), *Dictionary of scientific biography*, 3, pp. 272-278. Charles Scribners Sons, New York, 1971.
[4] '*Commant la science des nombres se peult appliquer aux mesures de geometrie*' (ms f.211^r).
[5] '*Commant la science des nombres se peult appliquer au fait de marchandise*' (ms f.264^r).

I call it the Triparty of Nicolas in the science of numbers, which was begun, continued and finished at Lyon on the Rhone in the year of salvation 1484.[6]

The history of the manuscript was given by Marre in his *Notice* in the *Bullettino*, where he writes:

> The manuscript . . . after having belonged, in all probability, . . . to Estienne de la Roche, known as Villefranche, was bought by an Italian gentleman named Leonardo de Villa, according to a note written in Latin on the *verso* of the last guard leaf at the beginning of the volume. Subsequently it became part of Colbert's library . . . then, on 11 September 1732, it passed from the Colbertine library into that of the King. . . . Today it bears the number 1346 in the *Fonds français* series at the Bibliothèque Nationale.[7]

In the catalogue of French manuscripts in the archives of the Bibliothèque Nationale published in 1868, the manuscript is described as including all the four sections referred to above, and there is no reason to doubt that it has always had these four sections since it left the hands of its author. The handwriting of all four parts is the same, and, despite the claim of Itard in the *Dictionary of scientific biography* that the manuscript is the work of a firm of copyists, it is now established, thanks to the work of M. L'Huillier, that it is in the hand of Chuquet. M. L'Huillier further describes an earlier version of part of the geometry, also in Chuquet's own hand and dating from about 1470.[8]

We know little about Chuquet himself. At the end of the *Triparty* he tells us only that he was a Parisian and a bachelor of medicine. Over his birth we can only speculate. Since he calls himself a Parisian, he was probably born in Paris, and most likely in the early 1440s, but we cannot state this with any confidence. Searches through the tax registers of Lyon reveal a certain *Maitre Nicolas, escripvain* in the registers from 1480 to 1484. He is said to have lived 'beyond the *Porte des Frères Mineurs* coming from the *rue de la Grenette* towards the *Muton*'.[9] In the registers of 1485 and 1487, he is given his full name *Nicolas Chu(e)quet* and described as being an *algoriste*, and in

---

[6] '*Je le nomme le Triparty de Nicolas en la science des nombres. Lequel fut comance medie et finy a lyon sus le Rosne Lan de salut 1484.*' (MS f.147^r; Marre, *op. cit.*, p. 814). Most of the translations used in this paper are taken from G. Flegg, C. Hay and B. Moss eds. (1984, copyright 1985), *Nicolas Chuquet, Renaissance mathematician. A study with extensive translations of Chuquet's mathematical manuscript completed in 1484*, Dordrecht, Reidel. This edition includes an appendix correlating the page numbers to the folios of the original manuscript and to the pages in Marre's edition in the *Bullettino* for the *Triparty* and problems, and in L'Huillier's edition for the geometry. For the above quotation, see p. 196.

[7] '*Le manuscrit . . . après avoir appartenu selon toute vraisemblance . . . à Estienne de la Roche, dit Villefranche, fut acheté par un gentilhomme italien, du nom de Leonardo de Villa, suivant une note écrit en Latin au verso du dernier feuillet de garde du commencement du volume. Il entra ensuite dans la Bibliothèque de Colbert, . . . puis de la Bibliothèque Colbertine il passa, le onze septembre 1732, dans celle du Roi . . . Aujourd'hui il porte le nombre 1346 du Fonds français de la Bibliothèque Nationale.*' Aristide Marre, 'Notice . . .', *op. cit.*, p. 580.

[8] L'Huillier, *op. cit.*, p. 81 ff.

[9] '*depuis la Porte des Freres Mineurs tirant par la rue de la Grenette vers le Muton*'. For this,

the registers of 1488 there is reference to *les hoirs Nicolas Chuquet, algoriste*. We can therefore conclude reasonably that he arrived in Lyon about 1480, and died in 1487 or 1488. Initially, it would seem that he earned his living a a copyist or writing-master. There seems to be no external evidence to substantiate Chuquet's claim to be a bachelor of medicine. His name does not appear in the records of the Faculty of Medicine in Paris, but we must note that the records for the period are incomplete, and it is even possible that he might have studied under a different name. As a bachelor of medicine, he would also have been a master of arts. It seems likely but is by no means certain that at some time he paid a visit to Italy. Internal evidence from his manuscript, and especially from the geometry, suggests Italian links. Marre suggested that there are a number of Italianisms in the *Triparty*,[10] but it is difficult to distinguish these from Latinisms. There was a large Italian colony in Lyon during the period of Chuquet's residence there. Internal evidence of Italian links is thus not necessarily evidence of an actual visit to Italy.

Other important information from the tax registers of Lyon relates to the de la Roche family, and in particular to Estienne de la Roche whose *Larismethique nouellement composee* was published in Lyon in 1520 and subsequently in a revised edition in 1538. This work, although known before the nineteenth century, came into special prominence in 1841 when Michel Chasles drew the attention of the members of the *Académie des Sciences* to its existence as the earliest treatise on algebra printed in France. Chasles did note, however, that de la Roche cited a work on algebra by Nicolas Chuquet, and he expressed the hope that this had not been forever lost. The discovery of Chuquet's manuscript less than forty years after Chasles' paper revealed that de la Roche, in entitling his work *Larismethique nouellement composee par maistre Estienne de la Roche*, was concealing the fact that a major part of it was little more than direct copying from Chuquet. At times this was not even particularly competent, since some of Chuquet's more valuable innovations were omitted. The Lyon tax registers establish that de la Roche's father lived in the Rue Neuve, close by the Rue de la Grenette, in the early 1480s, and that in 1493 Estienne de la Roche owned a house there as well as some property above Villefranche, and he is described as a qualified master of *argorisme*. De la Roche's plagiarism confirms that Chuquet's manuscript had come into his possession. Indeed, some marginal annotations are in de la Roche's hand. It seems likely that de la Roche may have been numbered amongst Chuquet's pupils.

---

and other information from the tax registers, see the preface to H.L'Huillier ed. (1979), *Nicolas Chuquet, la Géométrie. Première géométrie algébrique en langue française (1484)*, Paris, Vrin, and also his 'Eléments nouveaux pour la biographie de Nicolas Chuquet', *Revue d'histoire des sciences 19* (1976), pp. 347–350.

[10] Marre, 'Notice . . .', *op. cit.*, p. 585.

Chuquet is not very informative about his sources. He mentions Boethius and Campanus in the *Triparty*, and makes a reference to Archimedes as 'a certain wise man'. In the geometry, we find a reference to Lull. There is also, in the problems, criticism of the work of a certain maitre Berthelemy de Romans, also cited by Jehan Adam in 1475 as Bartholomew des Rōmanis.[11] Chuquet tells us that Berthelemy had formerly been a member of the preaching order at 'Valance' and a doctor of theology. The vernacular arithmetic which developed in France in the fourteenth and fifteenth centuries was influenced by material which had first appeared in Italy,[12] and Chuquet's work cannot have been an exception to this. These Italian works were inspired by the great thirteenth-century genius Leonardo of Pisa, whose *Liber abaci* appeared first in 1202 and again in a second version of 1228. There are many Italian arithmetics, dating from before Pacioli's *Summa de arithmetica geometria proportioni et proportionalita*, published in 1494—a work with which Chuquet's manuscript must necessarily be compared. Chuquet might have had access to these in Paris, or through the Italian community in Lyon, and possibly in Italy on a visit or visits for which we do not have concrete proof.

Of special interest is the question of whether or not Chuquet was familiar with the writings of Nicole Oresme, who, more than a century before Chuquet, expounded the concept of integral and fractional exponents. The earlier version of Chuquet's geometry is in a manuscript in the same hand as that of 1484, part of which is a copy of Oresme's treatise on the sphere. However, Oresme's writing on exponents occurs in his *Algorismus proportionum*, and we have no evidence that this particular work passed through Chuquet's hands. If it did, we have to face the question of why so daring an innovator as Chuquet failed to discuss fractional exponents.

Chuquet does not give a list of the contents of his manuscript, though he refers both to 'parts' and to 'chapters', and there are adequate headings to sections throughout the manuscript. The first part of the *Triparty* is an arithmetic covering roughly the same kind of ground as many of the early printed arithmetics and Italian abacus manuscripts: fractions (called by Chuquet *nombres routz*), progressions, perfect numbers, proportions, the rule of three, the rules of single and double false position, and methods for solving certain kinds of indeterminate problems. Chuquet also gives a rule, called *le rigle des nombres moyens*, or 'rule of intermediate numbers', of

---

[11] Problems 69 and 105 in the 'Appendice . . .', *op. cit.*, pp. 432 and 442; MS f.167^v and 186^v; Flegg *et al*, pp. 210 and 220-1. For Jehan Adam, see L. Thorndike,'Jehan Adam', *American mathematical monthly* 33 (1926) pp. 24-28, also his *Science and thought in the fifteenth century*, New York, Columbia University Press, 1929.

[12] Warren Van Egmond, *Practical mathematics in the Italian Renaissance. A catalog of Italian abbacus manuscripts and printed books to 1600*, Florence, Istituto e Museo di Storia della Scienza.

which he says he was 'at one time the inventor'.[13] This rule produces a fraction lying between two given fractions by adding the numerators and the denominators respectively. The rule of intermediate numbers is the one mathematical concept in all his manuscripts which Chuquet claims to be his own invention. In fact, the principle of this rule can be traced back to Greek times, though it may be doubted if Chuquet was aware of the relevant Greek sources.[14]

Chuquet makes use of negative numbers both as coefficients and as exponents. Also, he allows a negative solution of an equation, which he interprets as a debt. He calls a negative number *ung moins*, and there can be little doubt that he conceived negatives, along with roots, surds, and even zero, as numbers which could be manipulated in the same way as positive integers. His rules for basic arithmetic computations, including zero, go further than the tradition of Leonardo of Pisa. In having this very general concept of number, even if a practical rather than a theoretical one, he was very much in advance of his age.

With exponents, the most obvious thing is his notation, and especially his use of zero to represent the unknown quantity in an equation, as he says, 'without any denomination'. Thus $.12.^0$ represents our $12x^0$ (or the number 12). He thus clearly understands that any number raised to the power zero gives us unity. He provided all the necessary rules for calculating with the various powers of the unknown, calling this the *premier* rather than the *chose*, *cosa* or *res*. He gives many examples of products and quotients, such as: $.12.^3$ multiplied by $.10.^5$ giving $.120.^8$ (that is, $12x^3 \times 10x^5 = 120x^8$), and $.96.^3$ divided by $.6.^1$ giving $.16.^2$. More interesting examples of division are: $.72.^1$ divided by $.8.^3$ giving $.9.^{2.\bar{m}.}$, and $.84.^2$ divided by $.7.^{3.\bar{m}.}$ giving $.12.^5$. We can even find $.84.^0$ divided by $.7.^{0\bar{m}.}$ giving $.12.^0$. Chuquet chooses his calculations so that fractional indices are avoided, and he does not use his exponential notation with integers alone. Having called his unknown the *premier*, he then proceeds logically to refer to second, third, fourth numbers and so on. This is a much more convenient nomenclature than the older geometric one, as it avoids the complicated combinations of the adjectives 'square' and 'cubic'.

Claims that Chuquet invented logarithms are not substantiated by the manuscript. What we find is a table in which powers of two, headed *nombres*, are juxtaposed with their corresponding exponents, headed *denominacion*, the latter consisting of zero and the natural numbers. This is purely for the purpose of demonstrating from practical examples when it is appropriate to add or subtract exponents. We might at best call this a 'germ'

---

[13] '*Il y a aussi la rigle des nombres moyens de laquelle Jadiz Je fuz Inventeur par le moyen de laquelle jay fait aulcuns calcules que par deux posicions Je ne pouvoye faire.*' (MS f.83ʳ; Marre p. 736; Flegg *et al* p. 144.

[14] See the papers by Jacques Sesiano and Guy Beaujouan in this volume.

of the idea of logarithms, but no more, because the concept of expressing all numbers in terms of a common base is absent. In any case, neither the nomenclature 'second number' instead of 'square number' nor the association of successive powers with the natural numbers is new: the former can be traced back via the eleventh-century Byzantine writer Michael Psellus to the later Alexandrian period of Greek mathematics, and the latter possibly much earlier—perhaps to the time of Archimedes, as Itard suggests.[15]

In the solution of equations, Chuquet restricts himself to the types of equation which can be found in Hindu and Arabic works. In the case of quadratic equations, he does not shirk the question of two roots like some of his Western predecessors, such as John of Seville. Indeed, he explains:

when the R^2 [square root] of the remainder is added to the half of the intermediate (term), it produces a number. And when it is subtracted from it, it produces another number, both of which have the properties which it is necessary to have, and consequently one can take whichever one wishes.[16]

He then goes on to choose the example $.3.^2 \bar{p} .12.$ egaulx a $.9.^1$, and carries out the necessary computation only to discover that he is faced with complex roots, which he calls *impossible*. He also deals with simultaneous equations in two variables where one of these is of a form which enables him immediately to eliminate one of the variables. Such simultaneous equations had been known and solved from Babylonian times, and, although his notation permits the representation of only one unknown at a time, no notational problem arises in such cases. He also tackles certain higher-order equations, though these are of the specialized kind where the unknown appears only in the *m*th and *2m*th power. He admits his limitations, for the *Triparty* ends with these words:

There remains still for the perfection and completion of this book to find rules and general canons for three diversities of numbers unequally distant. And again for four or more diversities of numbers being equally or unequally distant the one from the other. These are left for those who would wish to proceed more deeply. And thus to the honour of the glorious Trinity . . .[17]

There is, however, one further point in regard to the equations which, following Itard,[18] it is important to note. In one problem, Chuquet escapes

[15] Itard, *op. cit.*, p. 274.
[16] 'quant la R^2 de la reste est adioustee a la moictie du moyen elle produyt ung nombre. Et quant elle en est soustraicte elle en p̄nte ung ault.^{e} qui tous deux ont les propetez quilz conuient auoir et pourtant peult on prandre lequel que lon veulx.' (MS f.139^r; Marre p. 804; Flegg *et al.* p. 191)
[17] 'Reste encores pour la perfection et acomplissement de ce liure trouuer rigles et canons generaulx pour troys differances de nombre inegalement distans. Et encor̄ pour quatre ou pluβs differances soient egalement ou inegalement distans lune de lautre. Lesquelles sont delaissees pour ceulx qui plus auant vouldrōt p̄funder. Et ainsi a lonneur de la glorieuse t'nite . . .' (MS f.147^r; Marre p. 814; Flegg *et al.* p. 196.
[18] Itard, *op. cit.*, p. 276.

Nicolas Chuquet—an introduction 71

mainly of worked examples; only occasionally do we find a general rule or a detailed discussion. Its title is:

How the science of numbers may be applied in matters of merchandise . . .[31]

The first chapter details the four rules of arithmetic; thereafter, most of the work consists simply of applications of the rule of three, though alternative methods of solution are occasionally given. Unlike some of the early printed arithmetics, there is no example of reckoning with tokens. It contains neither original nor advanced mathematics.

Chuquet's commercial arithmetic provides information about the economic life of fifteenth-century Lyon, about the currencies in use, the weights and measures, and about financial arrangements between partners in business enterprises. The various currencies mentioned come from different part of France, and from Germany and Italy, probably all to be found in the market at Lyon. Unlike the problems, where it is clear that Chuquet was primarily interested in applications of rules rather than in the reality of the problems themselves, in the commercial arithmetic we do find him providing examples intended to be reasonable. Thus, a problem on inflation, or one on the melting down of silver, may well be genuine evidence for the social or economic historian. It seems reasonable to infer from some examples, for instance, that exchange rates fluctuated considerably. Other examples may well relate to actual methods of counting used by money-changers. We know that the late fifteenth century was a period of rapid commercial growth in Lyon, a city in which Louis XI took a personal interest. He gave royal protection to the fairs established there in 1464, which made Lyon one of the most important markets of the Western world. The fairs in their turn stimulated the growth of banking, and by the end of the century Lyon had become an important centre of printing. All this commercial activity must have produced a demand for mathematical education in the vernacular. In some cases, all that would be required would be a fairly rudimentary knowledge of arithmetic together with its commercial applications. The last part of the manuscript provides precisely this. Its style is similar to that of the commercial arithmetics in use in Italy.

The absence of compound interest problems in the commercial arithmetic may appear strange, but it is clear that Chuquet wished to keep this particular part of his teaching elementary. Compound interest does occur in the problems on the *Triparty*, so it may well have been that Chuquet adapted his level to the needs and abilities of individual pupils, reserving his algebra and its applications for the more advanced.

In the light of the knowledge which we have today, we can give Chuquet

---

[31] '*Commant la science des nombres se peult appliquer au fait de marchandise*' (MS f.264^r; Flegg *et al* p. 296).

the title accorded for a time to de la Roche and call him 'the father of French algebra'. Like his contemporary Pacioli, he was both an abstract and a practical mathematician. The emphasis on rules and methods in the *Triparty* and the problems is nicely balanced by the practicality of the commercial arithmetic, and in the geometry he shows himself in both lights. He was not in the scholastic tradition, and his rules, though sometimes justified by analogy with more elementary arithmetic and copiously illustrated by examples, are not proved by either classical or modern standards.

It is perhaps interesting to speculate whether or not he believed that by means of *le rigle des nombres moyens* any rational number could be obtained. We may well wonder why that rule alone was claimed by him as being original. Is there perhaps somewhere and totally unknown to us a manuscript source for Chuquet's work which contains *le rigle des premiers*?

Much more work needs to be done in comparing Chuquet's manuscript with other contemporary, earlier, and later works, particularly those in French and Italian. There are still many fifteenth-century manuscripts in Italy which lie hardly examined on the shelves of library archives. There is a need also to look at works by those with whom Chuquet may have shared common sources, and to look in the works of the sixteenth and seventeenth centuries for the reappearance of what are regarded as Chuquet's innovations. Some of this work of comparison has been done, but much remains undone. Three examples, all of which suggest possible questions not adequately answered, are:

- The practice of underlining as a precursor of modern brackets occurs in the sixteenth-century writings of Alexander and Bombelli—a possible historical link here with Chuquet
- The term *premier* is found again in Stevin—perhaps something of significance here!
- Where can the next occurrence of a rational function such as

$$\frac{30 \cdot \overline{m} \cdot 1^1}{1^2 \cdot \overline{p} \cdot 1^1}$$

which is included in the *Triparty*, be found—and would this establish a link with the 'father of French algebra'

Chuquet's work is thus worthy of study today, not only for its content and the keen mind which that content reveals, but also for the many possible lines of research which it suggests—sociological and pedagogical as well as historical and mathematical.

The place of Nicolas Chuquet in a typology of fifteenth-century French arithmetics

GUY BEAUJOUAN

1. Introduction

In thinking about the place of Nicolas Chuquet in a typology of fifteenth-century French arithmetics, I understand the word 'typology' in the sense which has spread in recent years among historians due to the *Typologie des sources du Moyen Age occidental*, published by the Institut d'études médiévales of Louvain-la-Neuve.

In order to minimize matters of manuscripts, I refer to my paper 'Les arithmétiques françaises des XIVe et XVe siècles', given in 1956 to the 8th International Congress of the History of Science at Florence.[1] After various vicissitudes,[2] I reopened this file for the conference organized by R. Rashed in April 1984 at Marseille-Luminy, on the general theme of Mediterranean mathematics (until the very beginning of the seventeenth century). There I gave a report on 'la diffusion des arithmétiques commerciales en France aux XIVe et XVe siècles'.

Whereas, in my previous researches, my main endeavour was directed towards unearthing manuscripts, here I would like to stress what may illuminate the work of Chuquet. I should like to emphasize more especially the orientations, verifications and new investigations to which the conference of Marseille-Luminy inspired me.

2. The manuscripts

For those who have consulted numerous manuscripts, there is an inescapable contrast between university algorisms and merchants' arithmetics in the Italian style.

---

[1] Actes du VIIIe Congrès international d'histoire des sciences: Florence–Milan 1956 vol. I (Paris, Hermann, 1958) pp. 84–87.
[2] In particular, I had passed this file on for M. Paul Benoit's use; his contribution on Chuquet's commercial arithmetic is in the present volume. See also below, note 21.

- University algorisms, in Latin, were conservative, recopying Alexander of Villedieu or John of Sacrobosco, calculating on sand or dust with the possibility of erasure (avoiding therefore carryovers and thus lightening the appeal to memory). The setting of these texts shows that they were used by astronomers and astrologers.
- The merchants' arithmetics, in the Italian style, are quite different: composed in the vernacular, they teach calculation written with pen on paper, and they invoke commercial applications.

Among these French commercial arithmetics in the vernacular, it is possible to distinguish two principal groups. The first originates in the north of France, and is linked to the works of Jean Fusoris, the maker of astrolabes. The other, which is much more important, originates in southern and Mediterranean France.

3. Jean Fusoris and manuscripts from Northern France

The group from the north of France (Wallonia, Normandy, the Parisian region) is represented by the following manuscripts: Paris, Bibliothèque Nationale, **latin 7287** and **french 1339**; **Nantes 546**; Florence, Laurenziana, **plut. 29 cod 43**; **Tournai 86** (destroyed in 1940).

These arithmetics are linked with Jean Fusoris to the extent that, in the manuscripts, they may appear beside his treatise on the astrolabe, or beside a geometry which could be attributed to him. The disconcerting kinship of these texts can be re-examined with the aid of Emmanuel Poulle, *Un constructeur d'instruments astronomiques au XVe siècle, Jean Fusoris* (Paris, 1963).

A curious person this Fusoris: master of arts and of medicine, bachelor of theology, canon, but also the son of a pewterer who had taught him metalwork. His 'instruments of the seven planets' are among the most ingenious 'equatoria' of the Middle Ages. In order to sell them to the English, he rashly compromised himself. Thence, in 1415–1416, a trial for treason which today offers a quite exceptional type of information to the historian of science.

Often Fusoris had to sell the 'method of use' at the same time as the instruments themselves. It is thus that, between 1407 and 1412, probably in 1408, Fusoris composed for Pierre de Navarre, Comte de Mortain, a treatise in French on the uses of the astrolabe (with a third part of six chapters on the geometrical utilisations of the instrument).

In the manuscript **Paris Bibl. Nat. Fr. 1339**, for example, these geometric chapters are no longer to be found in the treatise on the astrolabe; they have been incorporated into the preceding practical geometry. Now Fusoris had already, in 1415, composed *Enigmata, id est ludos geometrie et astronomie*; surely, it concerns the same thing.

The place of Nicolas Chuquet in a typology of fifteenth-century French arithmetics 75

Since, in the manuscript **1339**, the arithmetic and the geometry are preceded by a prologue which seems to claim to be more or less common to the two treatises, it is tempting to link the arithmetic with Fusoris.

The manuscript **Paris Bibl. Nat. Lat. 7297** contains a précis of the Fusoris treatise on the astrolabe in its first version (with geometric uses). One might suppose that the arithmetic which accompanies it is a résumé of a 'Fusorian' version earlier than that of manuscript **1339**.

The Nantes copy, which is quite complete, is indeed affected by the influence of the southern arithmetics of which we shall speak in a moment.

In April 1984, at Marseilles, I was still hoping that it would be possible to

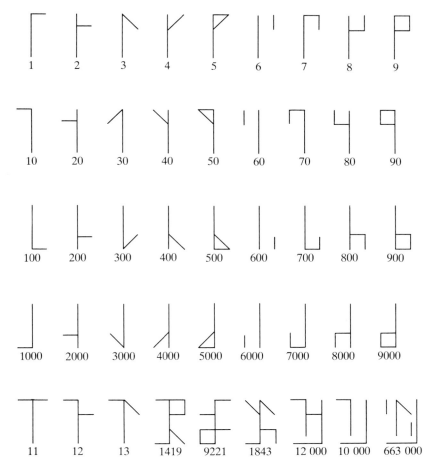

FIGURE 1.

recognize the original source of these arithmetics linked to Fusoris. For this I was counting on the examination of the Laurentian manuscript which I had not then seen. Unfortunately, comparing this manuscript **Firenze Laur. 29.43** with the manuscripts **Paris Bibl. Nat. Fr. 1339** and **Nantes 546** establishes that, in contrast with the geometry, which is quite stable, the arithmetics are extremely variable.

I had also placed some hope in the fact that the **1339** manuscript teaches a fairly rare 'Greek algorism'.[3]

The same numbers are to be found on the astrolabe known as that of Berselius;[4] the date of 1522 is when this astrolabe was given to a certain Hadrianus Amerocius. The writing and the astronomical data place the construction of the instrument at 1380 or thereabouts.[5]

The fair number of astrolabes which emerged from the workshop of Fusoris have the distinctive feature that on their rete, the star Cornu Arietis (the horn of Aries) has the correct declination in absolute value, but southern instead of northern. I should like to verify if this distinctive feature is to be found on the astrolabe of Berselius. I am, however, pessimistic, owing to the fact that the rete of the astrolabes from the Fusoris workshop comprise 21 or 22 stars,[6] whereas the astrolabe of Berselius has 29.[7]

It may be that these erudite considerations appear somewhat lengthy for such an uncertain result. The various arithmetics from the north of France, of which we shall shortly speak, have indeed a family resemblance. They are situated in the context of Fusoris, and hence in his wake, but without it being possible at present to ascribe the original paternity to him.

The important point is to recognize the chief characteristics of these texts of which, let us repeat, the most representative is that copied in **MS français 1339**.

- The operations (*especes* = 'species') are still 9 in number, as with Sacrobosco, with mediation being allowed to precede duplation.
- Calculation with counters is taught in parallel with calculation written with pen and ink. The arithmetic composed in Paris in 1475 by Jehan Adam is also wholly devoted to calculation with counters.[8] When, in

---

[3] Guy Beaujouan, 'Les soi-disant chiffres grecs ou chaldéens (XIIe–XVIe siècles)', *Revue d'histoire des sciences*, vol. 3, n°. 2 (April–June 1950), pp. 84–87; Jacques Sesiano, 'Un système artificiel de numération au moyen âge', typescript to be published in offset.

[4] Robert T. Gunther, *The astrolabes of the world* (Oxford, 1932) p. 349, n° 202.

[5] Sharon Gibbs, Janice A. Henderson and Derek de Solla Price, *A computerized checklist of Astrolabes* (November 1973), n° 202; Derek J. Price, 'An international checklist of astrolabes (first of two parts)', *Archives internationales d'histoire des sciences*, volume 8, n°. 32 (July to September 1955), pp. 243–263, especially p. 250.

[6] Emmanuel Poule, *Un constructeur d'instruments astronomiques au Xe siècle: Jean Fusoris* (Paris, Champion, 1963) p. 23.

[7] See above, note 4.

[8] Lynn Thorndike, 'The arithmetic of Jehan Adam', in his *Science and thought in the fifteenth century* (New York 1929; reprint Hafner, 1963), pp. 151–160 and 302–307.

1982, Michel Pastoureau prepared the section on counters for the *Typologie des sources du Moyen Age occidental*, a fact struck him and took him by surprise: a fair number of the oldest counters for calculation come, for example, from Nuremberg, Tournai, and Northern France; in contrast, there are practically none which originated with Italy, despite the importance of the *scuole* or *botteghe d'abbaco*. This observation does not appear to be due to haphazard preservation of counters. It reflects a reality, since, as Warren van Egmond has emphasized, the word *abbaco* designates practical arithmetic without any reference to the use of counters.[9] It does appear that, at the end of the Middle Ages, in Italy and in the South of France, calculation with counters was much less common than in the North.

- Certain problems take up the theme of the *Cautelae* which, in the manuscripts, serve as applications of Sacrobosco's algorism.[10] Other problems appear to be the adaptation of a Walloon model to the taste of Norman or Parisian users.[11] The applications are concerned with finance almost as much as with trade.

In brief, we have here a group of texts which adapt Sacrobosco to pen and ink calculations, but which only diffidently move away from this model.

4. Manuscripts from the South of France

Contrasted with this northern group is a southern group which is the more important because the new arithmetic appears to have been diffused in France by a kind of capillary process, starting from the southern regions.

But before presenting these texts it is worth emphasizing four points:

1) First of all, it should be noted that the *Liber abaci* of Leonardo of Pisa was unknown in France at the end of the Middle Ages. The *Quadripartitum numerorum* of Jean de Murs represents the only acknowledged exception to this. But only the end of the *Liber abaci* is used there, doubtless because Jean de Murs only read this text at the moment when he was preparing the

---

[9] Warren van Egmond, *Practical mathematics in the Italian Renaissance*. Supplement to *Annali dell'Istituto e Museo di Storia della Scienza* 1980, fasc. I (Florence, 1981), especially p. 6.

[10] Guy Beaujouan, 'L'enseignement de l'arithmétique élémentaire à l'Université de Paris aux XIIIe et XIVe siècles', *Homenaje a Millás Vallicrosa* col. I (Barcelona, C.S.I.C., 1954), pp. 93-124, especially pp. 115-123; Maximilian Curtze, 'Arithmetische Scherzaufgaben aus dem 14. Jahrhundert', *Bibliotheca Mathematica*, second series, 9 (1895) pp. 77-88.

[11] In the manuscript **Paris B.N. fr. 1339**, folio 59, there is a question about 'a man who makes a dress of cloth from Ypres (Ypres in western Flanders) or from Moutiervilliers (Montivilliers, in the *arrondissement* of le Havre) or from Rouen'. Another example which is even more characteristic is also on folio 59: 'Two men travel in one day between two towns at 50 leagues distance the one from the other, as one might say Rouen or Paris. And he who travels from Tournay or Rouen . . .'.

'questions' completing book III.[12] Thus, in the *Quadripartitum* of Jean de Murs, and therefore in the *Aggregatorium artis arismetrice* which is a plagiarism thereof, done in about 1425 by Roland l'Ecrivain,[13] the influence of Leonardo of Pisa is exerted only with regard to algebra, apart from a chapter discussing the composition of coins.

2) In the texts studied here, nothing betrays a direct and obvious plagiarism of the Italian models. Even in the case of Chuquet, A. Marre has greatly exaggerated in speaking of 'the large number of Italianisms which the *Triparty* contains.'[14]

3) The multiplicity of procedures proposed for the same operations betrays in general a real uncertainty, more than the concern to be complete.

4) These early vernacular arithmetics are difficult to classify. They are adaptations rather than copies. A large number of intermediate texts have disappeared. A manuscript that belonged to a merchant has, in fact, less chance of being preserved today than, for example, a volume which has always been kept in a monastery.

Let us now briefly mention the southern arithmetics which originated in Languedoc and Provence.

- the Pamiers algorism (ms. **Paris Bibl. Nat. nouv. acq.fr. 4140**)
- the ms. **Paris Bibl. Nat. fr. 2050**: text in French but clearly adapting an Occitan model
- the ms. **Nantes 546**: southern contributions related to the contents of ms. **fr. 2050** have been grafted onto a text from the North of France akin to that of ms. **fr. 1339** (in the 'Fusorian' family)
- the *Compendio de l'abaco* by a citizen of Nice, Francés Pellos (published in Turin in 1492)
- the *Suma de la art de arismetica* by Francesch Sanct Climent (published in Barcelona in 1482)
- finally, slightly aside from the others, the *Kadran aux marchans* begun in Bilbao in 1485 by the Marseilles merchant Jehan Certain

---

[12] Ghislaine L'Huillier, 'Regiomontanus et le *Quadripartitum numerorum* de Jean de Murs', *Revue d'histoire des sciences* 33 n° 3 (1980), pp. 193-214, especially p. 195. By the same historian: 'Le *Quadripartitum numerorum* de Jean de Murs', summarized in *Positions des théses de l'Ecole des Chartes: promotion 1979* (Paris, 1979), pp. 83-88.

[13] Thérèse Charmasson, 'L'*Arithmétique* de Roland l'Ecrivain et le *Quadripartitum numerorum* de Jean de Murs', *Revue d'histoire des sciences* 31 n° 2 (1978), pp. 173-176.

[14] Aristide Marre, 'Notice sur Nicolas Chuquet', *Bullettino di bibliografia . . . Boncompagni* 13 (1880) p. 585. As Jean Itard remarks, 'Upon examination, these so-called italianisms appear to be nothing more than latinisms.' (*Dictionary of scientific biography* 3, (New York, Scribner, 1971), article on Chuquet, pp. 272-278 especially p. 273). This article also appears in Jean Itard, *Essais d'histoire des mathématiques*, collected by Roshdi Rashed (Paris, Blanchard, 1984) pp. 169-179, especially p. 170 on the supposed italianisms of Chuquet.

A few brief observations on some of these texts may clarify what follows.

Of these treatises, the most important one for us now, because it is the one most closely related to the work of Chuquet, is indisputably the Pamiers algorism of ms. **Bibl. Nat. nouv. acq. fr. 4140**. The text is in Languedocian and the manuscript itself dates from the middle of the fifteenth century. But there are a number of signs which lead one to suspect that one or more earlier Occitan texts have been copied.

Several obvious errors of transcription are to be found, of which the most amusing is, at the beginning of the text, *'algorisme que foc natural de vida'*. The authors of the medieval West did not regard al-Khwārizmī as a sort of 'ecologist', but they usually thought of him as a native of India, like the numbers of which he was the principal propagator.[15] The faulty reading *vida* for *Inda* or *India* can only be explained by an Occitan or Spanish original. Moreover, according to Briquet's catalogue, the watermarks on the paper appear to be attested as from the years 1446–1448; nonetheless, as J. Sesiano has recently said, some of the problems appear to date from around 1430, insofar as one can judge this from various clues concerning coinage, in particular the currency of the coin known as the mouton.[16]

Another piece of evidence may be more insubstantial. On folio 118 (one of the flyleaves of a different kind of paper) there is a brief fragment on addition written in Occitan by the same hand as the notes on the *compotus* which refer to the year 1413 in the future tense. The numerical example in this fragment is found again on folio 20^v of the treatise itself. Perhaps it would be risky to regard this as flotsam from a lost Occitan treatise used by the Pamiers compiler.

Be that as it may, it is striking to note that among the elements appropriate to an Occitan commercial arithmetic, some passages are intermingled which include scientific terminology left in Latin and some arguments which exhibit, somewhat ponderously, knowledge with a flavour of the universities (the perfection of the numbers 3 and 6, preliminary considerations on the extraction of roots of fractional numbers, etc.). The algorist of Pamiers did not think he should attach his name to this work, but it is noteworthy that in the *Suma* of Barcelona, Francesch Sanct Climent states specifically that he had submitted his work both to a university teacher called Rapita and to a former Master of the Mint at Perpignan, Jachme Serra.

If we compare the arithmetic of Pamiers with the treatises from the north of France, the differences are clear:

---

[15] See, for example, what a glossarist of Sacrobosco's algorism writes (**Paris B.N. lat. 7420 B**, folio 52): '*[Algorismus] dicitur ab Algoro inventore Yndo existente.*'

[16] An article which has been published since the oral presentation of the present paper: Jacques Sesiano, 'Une arithmétique médiévale en langue provençale', *Centaurus* 27 (1984) pp. 26–75, in particular note 8.

- the traces of the influence of Sacrobosco are faint; for example, mediation and duplation are no longer the object of special chapters.
- no mention of calculation with counters
- the way of carrying out operations is in general more modern, and nearer to our own. Of all the operations, multiplication is, without doubt, the best test of the evolution of the fourteenth- and fifteenth-century arithmetics:[17] in the Pamiers algorism, it is carried out in the modern manner, but the multiplier is placed above the multiplicand and use is made of a grid; the method of the 'divided square' (the Italian *gelosia*) is evoked, but rapidly.
- application of the rules of three and of false positions are treated at length with numerous problems of which particular examples refer to Avignon, Perpignan, Montpellier, Béziers and Barcelona. Although one finds the so-called rule of apposition and remotion, the problems are more rarely recreational, and above all more clearly commercial, than in the family of texts loosely attached to Fusoris.

The manuscript **français 2050** was transcribed around 1460. It is in French, but it derives very closely from a model in Occitan. Words like *partideur* (divisor), *nombradeur* (numerator), *denombradeur* (denominator), *legue* (league) could only come from the Spanish or the Occitan. Certain passages betray a close kinship with the corresponding places in the Pamiers algorism, as is proved for example by the collation of texts on the rule of two false positions. Such a comparison conduces to thinking that the manuscript **2050** is in part the French adaptation of an Occitan treatise which was also used by the Pamiers compiler; this latter probably remained closer to the original.

Let us add some other remarks on this manuscript **2050**. Except in the prologue, which is very close to that already encountered in the ms. **B.N. lat. 7287**,[18] there is once again no trace of the influence of Sacrobosco. Taking as a test the case of multiplication, it may be observed that this operation is carried out in the modern manner (although the 'multiplying number' is

---

[17] For most operations, the procedures of calculation with erasure on sand can be easily transposed to pen and ink on paper. Formulae of the sort 'in place of the previously destroyed figure, set . . .' can be easily interpreted, 'cross out and write above'. Algorismic multiplication with 'anteriorigation' of the multiplier is the operation which is most difficult to transpose in this way; it is therefore a very revealing test of treatises on calculation with pen and ink. As for multiplication 'in square' (which the Italians called *gelosia* by analogy with the trellis of Venetian blinds), this was taught in France from the first half of the fourteenth century, in two treatises which were otherwise faithful to calculation with erasures: the *Algorismus minutiarum* of Jean de Lignères and the *Quadripartitum numerorum* of Jean de Murs.

[18] '*La premiere congnoissance de toutes choses naturelles a leur premiere rassine est venue et entree en entendement humain par raison de comptes et de nombre . . .*' ('The first knowledge of all natural things in their original root came to and entered human understanding by reason of calculations and number . . .') The same incipit, with minimal variants, in the ms. **Nantes 546**.

these modifications is clearly instructive about the paths of transmission—I was going to say the 'geography'—of calculation practices linked to economic life. It would also be helpful to check whether such adaptations to the local user took care to be in agreement with reality.

I come now to the questions concerning Nicolas Chuquet himself.

If, as we have attempted to show, the work of Nicolas Chuquet derives on the one hand from the common stem of Occitan and Provençal arithmetics of the fifteenth century, was it through the enigmatic Barthélemy of Valence (or Romans) that he had access to this tradition? Shall we come one day, if not to discover, at least to reconstruct the treatise of this mysterious Dominican?

As for his invention (limited after all) of the 'rule of intermediate numbers', Chuquet used the word *jadis* ('at one time', or 'formerly'). One may therefore wonder if, as H. L'Huillier has shown in the case of the *Géométrie*, the arithmetical part of the *Triparty* is not a revival of a nearlier composition. In reading the *Triparty* and its appendices attentively, one comes to surmise the juxtaposition of elements which belong to different strata; the homogenization of the language and of the content[28] is not always perfect.

The case of the *Géométrie* is clear, since its first composition has actually been preserved (Paris, Bibl. Nat. **Nouv. acq. fr. 1052**); according to H. L'Huillier[29], this first version reflected Chuquet's Parisian studies. On the other hand, the definitive text is impregnated by Italian influence and marked by the application of the 'rule of first terms'.

In admitting the hypothesis of an earlier version of the arithmetical part, one is led to think that it must originally have looked rather like the Pamiers treatise. When, in 1484, confident in his mastery of algebra, Nicolas Chuquet composed the definitive synthesis of his teaching, he wanted to make his *Triparty* a relatively skeletal and abstract work; he therefore transferred to an appendix certain elements of his early work (problems, commercial arithmetic), not without sometimes modifying the methods of solving various problems.

But the fundamental question remains that of the ultimate Italian source for Chuquet.

Besides the manuscript of Chuquet, Estienne de la Roche possessed a treatise of a certain Philippe Frescobaldi, a Florentine. This text does not appear in any bibliography, and it must have been a manuscript. One may wonder if Chuquet had not been the previous custodian and therefore the user.

In the manuscript **Firenze riccardiana 2241**, *La pratica della mercatura* of

---

[28] See the paper by M. L'Huillier in the present volume.
[29] Nicolas Chuquet, *La Géométrie*: introduction, text and notes by Hervé L'Huillier (Paris, Vrin, 1979).

Francesco Balducci Pegolotti (about 1340) has been intelligently copied *per mano di me Filippo di Niccolai Frescobaldi, in Firenze, questo di 19 di marzo 1471*.[30]

Although he has not yet published his discovery, I think that I may say that H. L'Huillier has discovered a Florentine banker called Philippe Frescobaldi who lived in Lyon from 1485 to 1487. It is a safe bet that this was the same person and that Chuquet knew him.

Was this the person who introduced Chuquet to the knowledge of algebra which was particularly prevalent in Tuscany? Must one, on the contrary, see in him only a specialist in commercial and financial practices? B. Moss has emphasized in this volume that in the matter of commercial practices, Estienne de la Roche has added to Chuquet.[31] If the source of these additions can be traced back to Philippe Frescobaldi, this would clearly bring him less glory than the honour of having taught algebra to Chuquet.

This conference has shown at the very least that in the history of mathematics, even in domains which appear to have been explored, there still remain things to seek and even to find.

---

[30] Francesco Balducci Pegolotti, *La Pratica della mercatura*, edited by Allan Evans (Cambridge, Mass., Mediaeval Academy of America, 1936), pp. XI–XIII, XLII and 383. I was indeed familiar with Pegolotti's text, but it was J. Sesiano who, after the conference, drew my attention to the role of Frescobaldi in the transmission of *La pratica della mercatura*.

[31] See, in this volume, her paper which endeavours to rehabilitate Estienne de la Roche.

Concerning the method employed by Nicolas Chuquet for the extraction of cube roots

HERVÉ L'HUILLIER

WHILE I was preparing my contribution to the Colloquium on Mediterranean mathematics which took place at Marseilles in April 1984, I became interested in some approximation formulae used in the Middle Ages and at the beginning of the sixteenth century to calculate the cube root of a number. Martin Levey presented some of these formulae in his work on Kushyar Ibn Labban and Hindu arithmetic. I should like to add some remarks to what he has written, and to compare these formulae with the method of Nicolas Chuquet, the fifth centenary of whose work on arithmetic and algebra we are celebrating in this conference.

1. The extraction of cube roots in the Middle Ages

Even if, at Marseilles,[1] I drew attention to the differences which may be observed in the way in which several medieval Western texts treated the process of extracting cube roots, the methods are all closely related, and proceed according to the route which I shall sketch briefly here.

The number whose root must be found is split into slices of three digits, starting from the right, just as the number is divided into slices of two digits when finding the square root.

Next, the calculation begins with the group on the left which may comprise one, two or three digits. Its integer cube root is sought; this is easy, because it is one of the integers between 1 and 9. After this, the cube of the number which has been found is subtracted from the group on the left.

The second stage is more complex. One is confronted with a number made up of the remainder left over from the first stage followed by the second complete group of three digits. It is necessary to find the root of this number, or, more precisely, the second figure of the root, because at this step the root will consist of two figures, and the first is already known. The second will

---

[1] On the occasion of the colloquium on mediterranean mathematics, 16–21 April 1984, in my paper: 'Le traitement du problème de l'extraction de la racine cubique par Nicolas Chuquet, Luca Pacioli et Frances Pellos (1484–1494).

90   Nicolas Chuquet and French mathematics

also be a number between 1 and 9. Following Chuquet, one proceeds by trial and error as follows: append the suspected number to the digit already found multiplied by 3, then once again by this number that has been found, then again by the suspected number, and finally by 10; to this product, add the cube of the suspected number, and subtract the total from the number with which the second stage began.

Example: $\sqrt[3]{4\ 913\ 087}$

Remainder: 3

$\underline{}$

$4\ 913\ 087$

Root: 1   7

$\begin{array}{r} 17 \\ \times\ \ 3 \\ \hline 51 \\ \times\ \ 7 \\ \hline 357 \\ 343\ldots 7^3 \\ \hline 3913 \end{array}$

In the following stages, the same scenario is reproduced as many times as there are groups of three digits.

Once the last group of three has been reached, there may still be a remainder, as in the example above. Three attitudes were possible in the Middle Ages: to state that there is a remainder, that in consequence the root is not perfect, while being interested only in the integer part (this is what Chuquet does); to try to express the result by a fraction in an approximate way, since decimal numbers were not known in the Middle Ages (this is the attitude of Fibonacci and al-Nasawi, and also of Kushyar Ibn Labban); or finally, some people tried to express this remainder by a sex agesimal fraction (as in the case of Jean de Murs.)[2]

Looking at this procedure in another way, I have indicated that the second, then the third, then the fourth, digits, and so on, are obtained by trial and error. After a certain point this operation becomes rather tedious. Now, one may accept that a simple approximative formula expressed in the form of a fraction should give these figures without insisting on the method which I have presented. Only Nicolas Chuquet teaches an easy method of finding these figures; we shall see below from what formula they derive.

2. Approximations from Heron to Cardan

2.1 The approximation of Heron of Alexandria

In the Metrica,[3] Heron of Alexandria seeks the cube root of 100. He therefore makes the following calculation:

---

[2] cf. Jean de Murs, *Quadripartitum numerorum*, 1.II ch. 23.
[3] *Heronos Alexandros Metrikon (Rationes dimetiendi)*, Teubner, pp. 176–9.

The commercial arithmetic of Nicolas Chuquet 101

- *To give currency and circulation to a foreign coinage according to the currency of the coinage of the country* (one problem for each precious metal).
- *How a deneral or pattern can be made for making alloys.*

The text ends with some conversion tables.

The plan of the work has already shown the importance accorded to commercial questions, an importance which is even more clearly evident when we turn to a study of the problems. As early as the first exercise concerning addition, the practical aspect is evident: it is a question of a sum in *livres*, *sous* and *deniers*. A simple count shows that 86.5 per cent of the problems considered concern merchants; exercises which are not characterized by immediate interest for commerce are extremely rare.[12]

A large part of these problems deal with prices: the calculation of the total price of a commodity of which the unit price is known, or the reverse, the price of the unit as a function of the total, calculation of the net price or of the profit.[13] All of these problems, which are the most numerous, appear especially familiar to us because they are those of our childhood arithmetics. Their concrete character, their pedagogic value, and their everyday utility have allotted to them a prime place in the treatises of elementary arithmetic, beyond even any professional end. But at this end of the fifteenth century, such a tradition did not yet exist and the examples chosen by Chuquet had a greater practical significance than they may have today. Moreover, certain exercises refer to commercial operations which are much less familiar in the twentieth century. Thus the author attaches a great importance to barter, but even more to partnerships and to questions concerning coinages and precious metals.

(1) under the title of 'barter and exchange of commodities' twelve problems are found regrouped which are developed at some length.[14] The first question is fairly simple: it is a matter of an exchange of products using coin simply as a unit of reckoning which allows evaluation of the products. The first exercise asks how many pieces of a commodity valued at 9

---

[12] We consider that there are 8 out of 223 such problems and exercises. It is very difficult to decide whether or not an exercise is useful to merchants. Here we have not included the questions bearing on pensions, on the price of bread or of fruit, nor on calculations of the date or on the division of taxes; by contrast the examples of additions, subtractions, multiplication and division dealing with coins have been counted with the commercial problems.

[13] These problems occur in the applications of the simple operations or of the rule of three. Certain problems bring in taxes, especially *la gabelle* (f° 278ᵛ), or the costs of transport (f° 279ʳ). In one instance Chuquet calculates profit as a percentage: '*Item quant la charge couste 40 l. que doit on revendre la l. pour gangner a raison de 25 pour cent*' (f° 277ᵛ).

[14] These twelve problems take up four folios, that is eight pages, whereas the nineteen problems bearing on the application of the rule of three to goods sold by the pound are contained in one folio.

livres per item must be exchanged against 'all the goods' of another merchant which are worth 60 *livres*.[15]

But the situation rapidly becomes complicated and the problems become more difficult, because the merchants who offer goods at a price evaluated in 'ready money', that is to say in cash, 'mark them up', that is they ask a higher price when it is a question of an exchange. One who offers a piece of fabric, of *camlet* at 7 *livres* 6 *deniers* in ready money wants to *mark it up* to 11 *livres* in the case of an exchange. These practices are well known; they appear in other commercial arithmetics in Italy under the name of *baratti*[16] and they reveal the difficulties of a commercial system which does not have enough cash and where banking practices remain—except in the case of the important Italian commerce—very inadequate, indeed non-existent. Nonetheless *the barter and exchange of commodities* allows Chuquet to call to mind all the possible cases in the hope that 'none will be deceived'[17] and he even presents cases where the result leads him to advise against the business.[18] This is a distinctly practical aspect which reappears in the chapters on partnerships.

(2) Chuquet is interested first of all in companies and in partnerships of merchants who hold their capital in common for a certain time.[19] The problems present companies with a very simple structure, most often with two merchants, sometimes three, and always without outside shareholders, contrary to the case in Italy.[20] The participants supply funds but may also, in certain cases, give their labour, 'their service', or furnish, for the duration of the operation, the use of real estate: a house serving as a shop. Thence proceeds a series of problems of which the simplest are concerned with the distribution of profits when the agreement has been respected by the parties, the division then being made in proportion to the sums placed in the business. As soon as evaluation of the labour intervenes, matters become complicated. They are still more complicated when the terms of the contract have not been fully respected; then it is necessary to evaluate in a suitable manner the price of 'service' or of the hire of real estate.

[15] f° 288^v.

[16] The term is current in the Provençal arithmetics and those in French from the South of France. The *Kadran aux marchans* speaks of *changes et baratz* (P. Benoit, *op. cit.*, p. 222), and F. Pellos, in the *Compendion de l'abbaco*, uses *cambiar aut baratar* (F. Pellos, *op. cit.*, p. 144).

[17] The expression recurs twice, on folios 289^v and 291^r.

[18] f° 290^r.

[19] The first example of a company which Chuquet presents is as follows: '*Troys marchans ont fait compagnie ensemble en fait de marchandise dont l'ung a mis 10 escus l'aultre 6 escus et le tiers a mys 8 escus et tout ensemble par aulcun temps ont trouvé 15 l. de gaing. Assavoir moult combien d'icellui gaing vient à chascun d'iceulx considéré la mise d'ung chascun*' (f° 279^v).

[20] The workings of Italian companies are explained in particular by Y. Renouard, *Les hommes d'affaire italiens au Moyen-Age*, Paris 1968. The importance of foreign shares in Florentine companies is discussed on pp. 156–157.

The commercial arithmetic of Nicolas Chuquet 103

A characteristic example is furnished by the problem which describes a company: a man should contribute 500 *livres* as well as his house and receive two-thirds of the profit, whereas the other, supplying only 400 *livres*, will have only a third. Now both only contribute 300 *livres*. Two solutions are then possible. The temporary transfer of the house is equivalent to 300 *livres* of investment, since the total of the sums invested amounts to 900 *livres* and 400 *livres* yield one-third of the profit; one can therefore remunerate it accordingly, or, since the house represents one third of the anticipated investment, 300 *livres* out of 900, count it as the quarter of the capital as may be. Chuquet rejects this solution because it does not seem to him to be equitable.[21]

Questions of law recur several times in these problems. At times they seem to be more important than their mathematical treatment. Thus, when it is a question of dividing the losses of a partnership, where the two merchants did not contribute the agreed sum but one of them furnished his work, three solutions are proposed, all of which are mathematically exact. Chuquet then leaves open the choice of the solution: 'each person may undertake the one which seems to him the most judicial'.[22] Since commercial law cannot answer the requirements, he sets forth propositions which are founded on common sense, but which rest on the results of calculation. However, this mathematical equity turns out to be inadequate. Such an attitude does not seem to me to be an intellectual game, but rather the fruit of the reflections of a man who teaches commercial practice. However, the gaps do not only come from the law; they can also be ascribed to the inadequacy of mathematics applied to trade.

When the partners leave their capital in the company for unequal periods, Chuquet offers a solution by the rule of three, which divides the profits in direct proportion to the time and the sum invested. He rejects it in affirming that he believes 'that merchants do not form such companies' and accounts for this by insisting that capital increases together with the profits it yields. Not only the 'gain'—the profit—should be taken into account, but also the 'gain on the gain'.[23] He thus sets problems of compound interest.

Companies link together merchants who place their money in a business; 'covenants and bonds' which link merchants and their factors establish different relations. As in the old Venetian *commenda*,[24] the capitalist merchant supplies a sum which the factor, the active party, should 'administer and use in commerce'. 'For and because of his service', the factor receives either a part of the profit expressed as a

---

[21] f° 282ʳ.
[22] f° 283ʳ.
[23] f° 280ᵛ.
[24] Y. Renouard, *op. cit.*, p. 63.

fraction, in the text always $\frac{2}{5}$, or the revenue from one part of the capital; in the example considered, a yield of 200 *livres* on the 700 invested by the capitalist. The two practices express the same reality; the factor is recompensed as a function of the profits of the enterprise.

As a function of these two types of remuneration, Chuquet develops examples which are complicated by the fact that each party supplies a supplementary sum. The factor only does this 'with the consent of his master'. The division of the profits is then more an affair of law than of mathematics, even if this law rests on a calculated division. The solution presented seems to be simple and professes to be equitable. The profit on the part which the merchant has added is always divided in the same way as the initial capital, whereas the factor, having provided capital and work, receives all the profit from that which he has supplied 'and to each justice is done'.[25] This is a very simple form of contract, which does not give the two parties interests in the financing supplied by the factor, differing in this point from the Italian *colleganza*.[26]

(3) Monetary questions have an even more important place than that of partnerships. As well as elementary operations, addition, subtraction, multiplication and division, dealing with monetary units or calculations of price, specific problems are set which depend on the value of the coinage. The use of money for accounts, most frequently in *livres*, *sous* and *deniers*, besides the actual money in coins, required that merchants knew how to carry out conversions from the one to the other.[27] The text touches upon this type of exercise very early, before turning to questions of exchange.[28] In a world where, beside the royal coinage, there always existed seignorial coinages where foreign specie were in fact current, a wide variety of coins were to be found in a town such as Lyon. Tradesmen had to know how to practise simple operations of exchange. The first problem, which is exemplary in the matter, consists in finding the value of 20 florins at 18 groats each expressed in *reals* at 30 *sous*.

Both the times and the place required such conversions. If at the end of the fifteenth century, the epoch of brutal monetary changes is past, the rate for money for accounts experienced a slow but certain devaluation which must be taken into consideration. Chuquet considers the question from the angle of pensions and loans.[29]

All these exercises, easy and close to actual concerns, represent a mere

25 f° 285^v.
26 Y. Renouard, *op. cit.*, p. 63.
27 f° 271^r, 'on doit à ung marchand la somme de 200 escus de 30 s. 3 d. la piece. Assavoir moult quantz livres de 20 s. on lui doit bailler'.
28 f° 271^v.
29 f° 272^v, 'Item l'on doit à ung homme 15s. 8 d. du temps que l'escu valait 27 s. 6 d. Assavoir moult combien on lui doit maintenant que l'escu vaut 30 s. 3 d.'.

trifle by comparison to the last part of the treatise which is wholly devoted to the 'nature of gold and silver'. Divided into two, it deals first of all with silver, the precious metal most frequently employed in exchanges, then with gold. The two chapters have comparable contents even if the part devoted to silver is longer.[30] In both cases, one must know how to handle units, to perform 'the division . . . of marks'. If, in the case of gold, the mark is always divided into 8 ounces and the ounce into 24 carats, the ounce of silver can be divided into grains, primes, seconds, and thirds, or in *esterlins*, *mailles*, *ferlins* and fractions of *ferlins*.[31] Chuquet distinguishes the Paris mark from the Court mark, which is lighter and is worth $\frac{131}{144}$ of the first. This greater complexity in connection with silver emerges even more clearly in considering the standard of metal used. In all the calculations, and in practice, the gold for reference is pure gold, defined by its standard of 24 carats. Silver presents more difficult cases: the standard is given in *deniers* of sterling standard and *grains* of sterling standard. Twelve *deniers* of sterling standard or of the official degree of purity correspond to pure silver or fine silver, but there is also the King's silver, the basic alloy for coiners, from 11 *deniers* 12 *grains*, that is 930 thousandths of the Court silver to 11 *deniers* 22 *grains*, that is 910 thousandths, and the Ash silver of 11 *deniers* 18 *grains*, that is 980 thousandths.[32]

Starting from these data there develops a whole series of exercises which correspond to genuine problems which arise for merchants: problems of conversion from one accounting system to another; Court mark to Paris mark; or Ash silver to King's silver. But the essential thing is to know how to work out the *fine silver*, that is to calculate the quantity of precious metal contained in an alloy of which one knows the weight and the standard. The result can be expressed in units of weight or in *sous*, *deniers* and *grains* of fine silver, which in fact are only the divisions of the mark in precious metal. The importance of this operation appears in the fact that 21 problems are devoted to it, 14 for silver, 7 for gold.[33] If the *fine silver* is known, it becomes possible to give the value of the alloy when the price of the silver mark, expressed in standard units, is known. Under the title 'of alloys', Nicolas Chuquet considers two series of important problems for those who trade with precious metals or simply have to bring their money to a mint. First of all what must be known is the quantity of base metal, of *tare*, which must be added to a

---

[30] The *nature of silver* goes from f° 296ᵛ to f° 310ʳ whereas gold has only folios 310ᵛ to 316ʳ. This difference is explained in part because the varieties of alloys of silver involved a multiplicity of problems but also because a number of explanations given for gold also applied for silver.
[31] The two sorts of division were usual in the Middle Ages.
[32] In the current state of research, it has not been possible to define Court silver, nor the Court mark which in the account given here does not correspond with either the Troy marc, or with the mark of the Papal Court, or with the mark of the Tower of London.
[33] The problems concerning fine silver are found in folios 297ᵛ to 302ʳ, those on fine gold in folios 310ᵛ to 311ᵛ.

noble metal, or in one case, withdrawn, in order to obtain a specified alloy, a *billon*. The first example chosen for silver consists in seeking the weight of copper to add to 10 marks of fine silver to obtain a 'billon' of 7 *deniers* of alloy.[34] The second question is to know the standard of an alloy which comes from smelting several lots of metal which are more or less fine.[35] The necessity of knowing the standard often involves recourse to assaying, a series of physicochemical operations to ascertain, with the aid of a specimen, the proportion of precious metal contained in an alloy. Nicolas Chuquet is not interested in the technical procedure, even if he appears to know it; by contrast, he devotes several exercises to the calculation which, starting from the result obtained from a specimen, enables the determination of the standard.[36]

The last part of each of the two chapters is concerned with the *deneral* and with the 'making of coinage'. By *deneral*, the author means the weight of the coinage which can be calculated, starting from the number of coins cut from a mark of gold or of silver. The 'making of coinage' is connected with this, since the problems deal with the number, the weight or the standard of the coins which the minter, and ultimately the prince, should obtain from the mark of precious metal while keeping a part for the expenses and for the seignorial privilege. The variety and the number of problems show the importance of these questions for the education of a French merchant at the end of the fifteenth century.[37]

Two very practical problems conclude these chapters: one is to calculate, with the aid of the weight of fine metal, the value of a foreign coin, the other to determine the 'deneral or pattern for trafficking in billons'. This phrase conceals the practice, which was not exactly legal but doubtless widespread, of calculating the weight above which it is lucrative to send coins to the foundry. The heaviest coins will be resmelted, while the lighter ones will be kept.[38]

At the end of the treatise Chuquet placed two tables which he comments on and elucidates by means of numerous examples. The first shows the 'fine silver', the quantity of gold or of silver contained in an alloy; the value of the alloy is also calculated. The second table converts fine silver into the *King's silver*.

34 f° 304ʳ.
35 f° 306ʳ, '*Item un changeur a 3 m. 7 onces 15 d. à 9 d. 8 grains d'aloy, item 5 m. 4 onces 12 d. à 7 d. 12 grains d'aloy, item 8 m. 1 once à 10 d. d'aloy, on demande se tous les billons estoient fondus ensemble a quant d'aloy ilz seroient*'.
36 f° 305ᵛ, '*Du jugement des essayz*'.
37 f° 307ᵛ, '*Du deneral et du fait des monnoyes*', nine problems concern silver; f° 314ᵛ, seven problems concern gold.
38 f° 309ᵛ, '*Commant l'on peult faire ung deneral ou patron pour trebucher et billonner monnoyes*'; the problem recurs for gold on f° 316ʳ, '*Et tant doit peser le patron pour trebucher celle monnoye car toutes les pieces qui peseront autant ou plus que le patron on les doit mectre à part pour fondre car il y gaing*'.

Does the economic content of the application 'of the science of numbers in matters of commerce' correspond to the requirements of the French merchants, and in particular to those from Lyon at the end of the fifteenth century? The answer to this question will enable us to delineate the nature of Nicolas Chuquet's work.

Certain features stand out quite clearly. The commercial techniques taught were rudimentary by comparison with those which were then known in Italy: nothing on the drawing of exchanges, nothing on book-keeping, when, in this same period, Luca Pacioli presented double-entry in his *Summa de arithmetica*.[39] The question of compound interest appears twice, but it is considered in only one problem. A juridical question is set which exposes the gaps in French commercial law at the dawn of the Renaissance: Chuquet can only rely on general considerations, and on common sense, to try to disentangle what seems to him to be just. There also Italy possessed, both with respect to commercial techniques as well as with respect to law, an ample headstart which was also to be found in the mathematical solutions of the problems.

Although the commercial practices presented by Nicolas Chuquet on calculations of prices, partnership contracts, divisions of profits and losses, and loans were backward by comparison with those of the Florentines and the Venetians, nonetheless they were indicative of French commerce at the end of the Middle Ages,[40] and indicative as well of the importance and nature of monetary problems. Because the monetary system of that time was founded on a real coinage of gold and silver, because coins of all sorts circulated, because a standard currency was lacking, and the methods of substituting for it were still very inadequate,[41] because the era still lived under the influence of the changes that had stamped the fourteenth and fifteenth centuries, it was indispensable for every merchant to master the problem of coinage, to know which pieces should be kept and which pieces should be resmelted. Every merchant then was at times a money changer, and if trade in precious metals

---

[39] L. Pacioli, *Summa de arithmetica, geometria, proportioni et proportionalita*, Venice, 1494. The chapter *De computis et scripturis* runs from folio 197^v to folio 210^v.

[40] This retardation of French commercial techniques by comparison with Italy is clearly brought out by J. Heers, *L'Occident aux XIVe et XVe siècles. Aspects économiques et sociaux*, Paris, 1973, p. 214, *'Face aux cités d'Italie, aux ports d'Angleterre et de Flandre où le commerce de l'argent, les ventes à terms, les multiples formes d'emprunts connaissent une grande faveur, d'autres pays connaissent un retard considérable. Ceci surtout en France, sur la façade atlantique, à Paris, s'accompagne d'une grande faiblesse des moyens financiers.'*

[41] The mediocrity of the French bank is emphasized, in the case of Paris, by J. Favier *Nouvelle histoire de Paris. Paris au XVe siècle, 1380–1500*, Paris 1974, p. 368, *'La pratique du dépôt, par exemple, ne répondait à Paris qu'à des fins élémentaires: il ne s'agissait pas d'accélérer la circulation financière, de faciliter les virements et de pallier l'insuffisance de numéraire, mais simplement de garantir des obligations et d'user des coffres, réputés sûrs, des maniers d'argent professionnels . . . Dans le domaine du crédit, c'est le simple prêt qui l'emportait sur la place de Paris'*, and p. 370, *'La pratique du change tiré comme instrument de crédit est demeuré, à Paris, circonscrite dans un milieu étroit, celui des Lombards . . . '*.

no longer had the appeal which it had possessed at the beginning of the fourteenth century, it had nonetheless a place in the practices of any merchant who was careful in his business.[42] The place given to these problems locates the commercial part of Nicolas Chuquet's work in the line of descent of distinctively commercial arithmetics such as the **ms français 2050** of the Bibliothèque Nationale of Paris, or the *Kadran aux marchans*.[43]

Does Nicolas Chuquet's commercial arithmetic, adapted as it was to the economic and technical level of French commerce at the time, possess any specifically Lyonnais traits as well? The commercial techniques mentioned do not allow a definitive answer. By contrast the terminology, the units of measure, the products and the coinages may furnish some indications. The name for the sum charged to a person at the time of collecting a tax in the text is *vaillant*, a habitual term in Lyonnais fiscal matters.[44] Lengths are measured in *aunes* as in Lyon, but also in Paris, whereas the southern arithmetics used the *canne*.[45] Among the coinages, the *écu* and the *florin* are very much dominant, but there also appear local or regional specie such as the florin of Savoy or the *Lyon*.[46] In articles, remonstrances and memoranda addressed to the king between 1484 and 1487 by the highest authorities of the city, asking for the re-establishment of the fairs, the essential objects of trade in Lyon were above all cloth, then linen, wool, furs, silver, saffron and pins. Now, if there is nothing astonishing in Chuquet's text putting in the foreground cloth, and then linen, the importance of spices is to be noted, and trade with pins is the subject of a problem.[47]

The form of Nicolas Chuquet's commercial arithmetic is marked by the Lyonnais context, and its technical aspects are in keeping with the world of French commerce at the end of the fifteenth century. It corresponds to what an *escrivain* or an *algoriste* might teach a young merchant. The Italian influence appears to have been very weak; it may be possible to suspect it in the case of a loan with interest—the hypothesis is interesting but fragile. This simplicity and the elementary approach to commercial methods, compared to the transalpine model, recur in the domain of mathematics.

---

[42] On the money changers, their role and their decline in French economy and society at the end of the Middle Ages, see B. Chevalier, 'Les changeurs en France dans la première moitié du XIVe siècle, *Economies et sociétés au Moyen-Age*, Paris, 19, pp. 153–160.

[43] P. Benoit, 'La formation . . .', *op. cit.*, p. 223.

[44] Taxation of the Lyonnais is discussed in J. Deniau, *Les nommées des habitants de Lyon en 146*, Lyon and Paris, 1930, and in the edition by E. Philipon and C. Perrat of the *Livre du vaillant de Lyon en 1388*, Lyon, 1927.

[45] In **ms français 2050** of the Bibliothèque Nationale de Paris, the *canne* is the standard measurement used for textiles. One problem even deals with converting *aunes* of Lyon into *cannes* of Avignon.

[46] f° 271^v, '*l'on demande 498 florins 13 gros 7 d. de 18 gros 2/3 la piece quantz lyons de 27 gros 1/2*'; f° 272^v, '*l'on demande 46 florins de 12 gros la piece monnoye de Savoie comptant 8 fors pour gros quantz florins valent de 12 gros et le gros de 15 deniers . . .*'.

[47] M. Bresard, *Les foires de Lyon aux XVe et XVIe siècles*, Paris, 1914, p. 196.

The commercial arithmetic of Nicolas Chuquet 109

The mathematical content of the work is meagre. The solution of many of the problems depends on elementary operations of addition, subtraction, multiplication or of division. These exercises are complicated by the fact that medieval monies were expressed in complicated numbers, and the use of fractions occurs constantly. But the rule of three prevails to an overwhelming extent: the simple rule of three alone resolves almost half the problems. The theme of the *Triparty*, which defined it as 'the lady and mistress of the proportions of numbers',[48] is thus confirmed. The mathematical foundations of the commercial part, which is therefore indeed an 'application', all appear in the *Triparty* with the exception of the compound rule of three.[49] It is used essentially to calculate interests as a function of time or to accomplish exchanges between multiple coinages; it is on this subject that a classic problem of the arithmetics of the time is set:

... if 4 Paris *deniers* are worth 5 Lyon *deniers*, and 10 Lyon *deniers* are worth 12 Geneva *deniers*, one asks how many Paris *deniers* are worth 8 Geneva *deniers*.

The fact that a new rule is expounded perhaps explains that certain problems, which are presented by way of example, and considered to be necessary, are without direct relationship to commercial practices.[50] Chuquet only appeals once to the algebraic results which he developed in the *Triparty*. Discontented with the solution to a problem of interest by the compound rule of three, which does not take account of the 'profit on the profit', that is, of compound interest, he proposes another one which appeals to the *rule of first terms*, in other words, to algebra. The procedure is difficult to follow, but the result is exact.[51]

Other than this case, the problems remain simple and only require a limited mathematical education, quite distinct from the requirements of the *Triparty*. Nonetheless, the author makes great efforts to find the simplest

---

[48] *Triparty* p. 736.
[49] f° 292ᵛ, '*Comment les troys nombres de la rigle de troys se peuvent joindre et composer en plusieurs et diverses manieres*'.
[50] In at least one case the problem has no value for the merchant, even though it is posed in monetary terms: '*Quant 4 poyres valent 5 pommes et 7 pommes valent 13 noix et 25 noix valent 1 d. l'on demande quantes poyres valent 4 d.*' (f° 296ʳ).
[51] f° 293ʳ to 294ʳ, '... *Quant 100 escus en 12 moys gangnent 15 l. l'on demande que gangneront 60 escus en 8 moys* ... Response: *regardons que doivent gangner 60 escus en 12 moys c'est assavoir en ung an en disant si 100 escus gangnent 15 l. que gangneront 60 escus, multiplie et partiz si trouveras 9 l. et tant gangneront les 60 escus en ung an. Encores mectons que l'escu vaille 35 s. et le l. 20 s. ainsi les 9 livres vauldront 5 escus 1/7 qui adjoustez avec les 60 escus sont 65 escus 1/7 maintenant pour tant que 8 moys sont 2/3 d'un pour telle cause il convient sacher entre 60 escus et 65 1/7 deux nombres proporcionaulx par qu'il est demonstré ci devant en l'application speciale de la rigle des premiers en l'invention des nombres proporcionaulx et des deux nombres prend le majeur qui est R³ 254615 25/49 duquel lyeves 60 escus restent R³ 254615 25/49 d'escu moins 60 escus laquelle reste est moins de 6 livres et tant gangneront les 60 escus en 8 moys*'. Translating this into current notation gives $\sqrt{254615.5102}$, from which, when 60 is subtracted, there remains what corresponds to the value of the compound interest calculated by the usual twentieth century method.

solutions to these problems: he develops a complete pedagogy. The style of presentation varies according to whether the mathematical foundations have been studied in the *Triparty* or not. In the case where they have been, as with the elementary operations or the simple rule of three, Chuquet considers them known and is happy to set problems and to give their solutions. He passes from the simpler to the more complex; thus, in studying exchanges of commodities, he starts from an exercise which requires only a simple division, to end with a problem which requires two additions before arriving at a rule of three.[52] By contrast, when he presents a style of working which he has not previously treated, whether this be the compound rule of three or a more technical question such as the application of the rule of three to companies or the way of making 'a *sol* of fine silver', he first states the rule and then passes to the applications. This style of working, without any proofs, these statements of rules, quite different from those which we have today since they indicate a procedure to follow and not formal reasoning, is in keeping with the habits of the time.[53]

What is even more striking in the course of an attentive reading of the work is the constant concern to put within everybody's reach operations which appear to us quite simple. Chuquet is aware of this, and offers introductions to certain practices, such as:

... and who would wish to avoid the subtleties of fractions ...

and

... nonetheless, not everyone is skilful and expert in the matter of reckoning. . . .[54]

Thus, in several cases, in order to avoid multiplications by fractions, a source of possible errors, he uses successive mediations. In order to multiply by $\frac{3}{4}$ he sums two mediations, the first of which divides the initial number and the second which is applied to the first result; thus he adds $\frac{1}{2}$ and $\frac{1}{4}$.[55] When he presents the division of sums in money for accounts, a delicate operation, heal ways proposes, among the solutions which are retained, to convert the

---

[52] f° 290ᵛ and 291ʳ.

[53] At the head of the examples, the third case of the compound rule of three is defined thus: '*En la tierce partie de la règle de troys composée le premier nombre et le derrenier doivent tousjours estre dissemblans et le demande doit estre du derrenier nombre au premier et y peult y avoir 5 nombres ou tant que l'on veult dont d'icelle partie en est une rigle qui est telle. Multiplie ce que veulx scavoir par toutes les choses valant et puis partiz par la multiplicacion des choses valués*' (folio 295ʳ); or again the rule for alloys of metals: '*Pour alloyer tout billon et mectre en tel aloy que l'on vouldra ils en sont deux rigles generales dont le premiere enseigne d'aloyer deux especes de billon dont l'ung est terminé et l'autre indeterminé . . . La premiere rigle. De la faulte loy ou surloy du billon indeterminé faiz le second nombre et de tout le poyz d'icellui billon fait le tiers nombre*' (folio 304ʳ).

[54] f° 280ʳ and f° 307ʳ.

[55] f° 268ʳ, '*Ou aultrement prens les 3/4 de 546 8/9 en ceste maniere, prens pour le premier la moictié de 546 8/9 et encores la moictié de la moictié et auras en tout 410 escus 1/6 . . .*'.

sum into *deniers* first of all, before passing on to division.⁵⁶ Likewise, in certain cases he recommends simplification of the rule of three or of fractions.⁵⁷ But above all, several times he states *notables*, that is, 'simple and short rules' which should enable a number of calculations to be avoided. These *notables* appear from the beginning of the treatise in order to identify, expressed in *livres*, the value of a sum counted in subsidiary coins. They are particularly numerous in the chapters concerning gold and silver, and they assist in the calculations bearing on the standard of alloys in particular.⁵⁸

Finally, an extreme case: Chuquet even suggests, in order to establish the standard of an alloy after an assay, a solution which allows the result of the operation to be known by simple weighing.⁵⁹ The explanations which he gives are much more plentiful than when it was a matter of using the *rule of first terms* in order to calculate compound interest. Everything here suggests that he was developing an example which he usually presented to a public unfamiliar with even the simplest mathematics.

Except when required, Nicolas Chuquet thus refrains from appealing to mathematical knowledge beyond the level of the rule of three which he developed in the *Triparty*. By contrast, he displays a constant pedagogical concern. These facts, taken together with the technical character of the problems propounded, set his work in the tradition of arithmetic treatises intended for merchants.

Beyond this elementary truth, it is necessary to rediscover the tradition to which it can be related, the sources on which it may have drawn, and finally, to define what may constitute its originality.

Comparison with the contemporary *Novel opera de arithmetica* of Pietro Borghi, published in 1484,⁶⁰ shows clear resemblances: predominance of the rule of three, problems with companies, and barter (*baratti*), of precious metals and of alloys (*ligar metalli*). Among other things, the problem recurs, in the chapter on the compound rule of three, of the exchange of coins

---

⁵⁶ f° 270ʳ among other examples.

⁵⁷ Simplifications of fractions are frequent, for example in folio 270ʳ: '*155/310 qui abreviez sont 1/2*'. Simplification of the rule of three is only used once, and that in a very special case, to avoid considerable calculations with fractions (f° 280ʳ); it should therefore be judged that the practice of simplifying in the rule of three is not usual.

⁵⁸ f° 272ᵛ and 272ʳ, '*Notables sur le fait des monnaies. Les 60 quarnaires d'une chascune monnoye comptée de 4 en 4 pieces ou les 80 ternaires comptez de 3 en 3 valent autant comme l'une des pieces comtées vaut de deniers. Et par ceste rigle les 60 quarnaires de lyars ou les 80 ternaires d'iceulx valent 3 l. car ung lyart vaut 3 d.*'. This is equivalent to saying that the overall number of 4 coins multiplied by 60 or that of 3 coins multiplied by 80 is worth as many livres as the coin is worth deniers, indeed that the livre is worth 240 deniers, either 60×4 or 80×3. Here since the liard is worth 3 deniers, the sum obtained is $4 \times 60 \times 3 = 720$ deniers or 3 livres. Such a procedure would limit the errors of counting and calculation by avoiding any division.

⁵⁹ f° 307ʳ, '*Pourtant que chascun n'est pas habile ne expert en fait de comptes a esté trouvé une maniere de faire certains poyz representant le marc . . . et l'appellent aulcuns semelle ou poyz d'essay . . .*'.

⁶⁰ P. Borghi, *Novel opera de arithmetica*, Venice, 1484.

112 *Nicolas Chuquet and French mathematics*

between several places, with only the names of the towns, these coins and the numbers changed.[61]

But such a problem, which combines the features common to the commercial arithmetics of Chuquet and of Borghi, also appeared in the treatises in French in the second half of the fifteenth century, prior to the *Triparty*. These works all taught calculation with pen and ink, all set and resolved a number of problems useful to merchants, but they derived from two different traditions.

The first, which is Parisian and Norman, is strongly marked by the legacy of the university arithmetics descended from Sacrobosco;[62] they give space to calculation with counters, which was always used by public book-keepers and often accompanied geometries and treatises on the astrolabe.[63]

The second, represented chiefly by southern manuscripts such as the **Ms français 2050** of the Bibliothèque Nationale of Paris or the *Kadran aux marchans*, is not unlike the Occitanean texts and bears a more commercial but also a more innovative character. In these texts, calculation with counters disappeared as well as terminology and practices considered to be useless,[64] whereas monetary questions took a very important place. In the same fashion, Chuquet discarded the old traditions, and only taught written calculation on paper, while preserving the intermediate results, and in the same spirit he devotes long sections to 'the making of gold and silver'.

In these circumstances, should one seek for the sources of 'the application of the science of numbers to matters of commerce' in a simplified Italian tradition or in a Franco-Provençal tradition which was transmitted in particular by Berthelemy de Romans of whom Chuquet acknowledges himself to have been the disciple? In these terms, the question is not easily resolved; it is doubtless more complex. The vocabulary used in the commercial arithmetic, as elsewhere in the *Triparty*, does not reveal influences from either the South of France or Italy, contrary to A. Marre's view. The terms used, even if they possess a certain originality and reject the yoke imposed by the

---

[61] Of the problems which Chuquet sets (f° 245^v), one deals with Paris, Lyon and Geneva, the other with Flanders, Paris and Burgundy, whereas **ms français 2050** of the Bibliothèque Nationale de Paris cites the coins of Perpignan, Montpellier and Avignon in a problem which had already appeared in the Provençal text of the manuscript **nouvelles acquisitions françaises 4140** of the Bibliothèque Nationale (G. Beaujouan, *Recherches sur l'histoire de l'Arithmétique*, Thesis of the Ecole des Chartes, typescript, Paris 1947, p. 235). Borghi for his part mentions Modena, Venice, Corfu and Negropont.

[62] See G. Beaujouan, 'Les arithmétiques en langue française . . . ', *op. cit.*, and for the latter place P. Benoit, 'Recherches sur le vocabulaire . . . ', *op. cit.*, pp. 78–79. Sacrobosco's influence is apparent especially in the vocabulary and in the special place given to mediation and duplation.

[63] Two texts are characteristic of this tradition, **ms français 1339** of the Bibliothèque Nationale de Paris, and **ms 456** of the Bibliothèque Municipale de Nantes.

[64] Neither the *Kadran aux marchans* nor the **ms français 2050** of the Bibliothèque Nationale de Paris deals with mediation or duplation; this latter text ends with a list of currencies containing 161 coins with their description and their standard.

nomenclature of Sacrobosco, stem indisputably from the terminology of the texts from the North of France.[65] This remark would tend to support the hypothesis that Nicolas Chuquet had his initial training in Paris, in the university milieu.[66]

But, as an original mind, and faced with the problems of the Lyonnais commercial milieu, in contact with the southern, Provençal, or Italian world, he would have modified the language learnt in Paris at the same time as he would have omitted, in imitation of his southern models, all that appeared to him irrelevant to merchants, while he would have developed the very technical part dedicated to money.

In his definitive version, such as we have in the manuscript of the *Triparty*, the plan of the commercial arithmetic is clear and as a whole well structured, but the details of the organization reveal a number of perplexities. The chapter concerning the compound rule of three, defined as a function of mathematical criteria, follows the application of the rule of three to commerce (companies, agreements of merchants with their factors, barter) and precedes the questions bearing on coinage, all of which are chapters defined as a function of practical criteria. This lack of rigour in detail, the addition of material at the end of certain chapters, and above all the extreme unevenness of the difficulty of the problems and of the solutions indicated, tend to lead one to think that the elements which constitute the definitive construction of the work emerged from different moments of Nicolas Chuquet's thought.[67]

Apart from these remarks, the indications which would allow achronology to be attempted are sparse. In certain cases, the value of coins or the prices of precious metals may provide indications, but they are to be used with caution. As in textbooks of all times, the author could choose plausible but fictional values for the values of coins and of products. On the other hand, the market rate for foreign coins varied, and as well as the legal rate for precious metals, the price of purchase in the mint, there existed a commercial rate. Even the royal coins were sometimes changed at a parallel rate which was different from that fixed by the regulations. Thus it is impossible in the present instance to rely on the different values of the florin or the price of precious metals.[68]

---

[65] P. Benoit, 'Recherches sur le vocabulaire . . .', *op. cit.*, p. 87.

[66] H. L'Huillier, 'Eléments nouveaux . . .', *op. cit.*, p. 349.

[67] Thus the chapters dealing with companies and the associations of merchants with their factors are especialy developed. The problems are accompanied by remarks of a juridical character which gives them a different character from the rest of the work. One may ask if, in the chapter devoted to the application of the rule of three to merchandise, the paragraphs entitled '*Aultres raisons en fait de marchandises*', which contain the longest problems, were not added after the event (f° 278^r to 279 ^{r}). Other additions of this type are possible.

[68] On folio 265^v the florin is worth $18\frac{1}{3}$ groschen; on folio 271^r $18\frac{2}{3}$ groschen; on folio 272^v, florins at $10\frac{2}{3}$ groschen each are changed into florins at 12 groschen. These differences may be explained by the variation of the commercial rate of the Florentine florin but also by the multiplicity of coins of the florin type.

Only the information provided by the variations in the rate of the *écu*, a coin of royal gold, has a certain significance. In the majority of cases, the *écu* was worth 30 *sols* 3 *deniers*, which corresponds to the legal rate for the *écu couronne* between the 4th of January 1474 and the 2nd of November 1475.[69] A problem dealing with a loan even specifies that it is a matter of the current value, as against the old value of 27 *s*. 6 *d*. However, the same coin is quoted one time at 35 *s*. and another time at 35 *s*. $\frac{3}{4}$, which are values that do not correspond to any official exchange rate, since the *écu* was current at 33 *s*. the 2nd of November 1475 and then at 36 *s*. 3 *d*. the 11th of September 1483. All these data are found at the beginning of the treatise: the *écu* at 30 *s*. 3 *d*. appears as early as the chapters on addition and subtraction, whereas the application of the rule of three to 'matters of money' cites *écus* at 30 *s*., 30 *s*. 3 *d*., 35 *s*. and 35 *s*. $\frac{3}{4}$.[70] The legal currency of the *écu* at the time when Chuquet finished the *Triparty*, 36 *s*. 3 *d*., is never used. One may well think that a part of the work, which contains simple problems, comparable to those which are found in other arithmetics in French at this time, was composed between 1474 and 1475; supplements would have been added in the following years. The text would therefore have been begun at least ten years before the definitive composition of the *Triparty*.

Little by little, in the midst of uncertainties and gaps, there appear some indications which make it possible to propose hypotheses as to the origins and the elaboration of Nicolas Chuquet's commercial arithmetic. Brought up in the Parisian university setting, he would have been able to become acquainted with the practical arithmetics from the North of France very early but, teaching arithmetic to young Lyonnais merchants, he would have found in the southern treatises a more technical tradition, closer to the requirements of commerce which would beat the origin of the long chapters dedicated 'to matters of gold and silver' in particular. The first phase in the composition of the commercial arithmetic would date from the years 1474–5, before there is evidence of Chuquet's presence in Lyon. From 1475 to 1484 the text would be enriched and the examples would be multiplied. In contact with the Italians in Lyon, and doubtless also thanks to a trip to Italy, Chuquet would then have deepened his economic thought, just as he would have developed his mathematical thought. From this epoch would date the treatment of the problem of compound interests. Finally, at the time of the definitive composition of the *Triparty*, he would have been constrained, while making use of previously composed fragments, to add the chapter on the compound rule of three, having doubtless not considered it of sufficient interest to appear in the purely mathematical part, whereas he thought it was indispensable to merchants.

---

69 E. Fournial, *Histoire monétaire de l'occident médiéval*, Paris 1979, p. 136.
70 f° 273^r and ^{v}.

Thus the roots of the commercial arithmetic of Nicolas Chuquet are deeply embedded in the tradition of the commercial arithmetics of the end of the Middle Ages. The problems covered are those of French commerce in the fifteenth century, the methods used to solve them, the rule of three and crafts of calculation, were to be found in general in contemporary works; they were characteristic of the training of French merchants of the time. *The application of the science of numbers to matters of commerce* is more a technical than a mathematical text, and one intended, on all the evidence, to train merchants.[71]

I think therefore that I have answered, at least in part, two of the questions raised in the introduction. It seems to be much more difficult to provide an answer to the third which asked if it were possible to establish connections between the commercial arithmetic and the rest of the *Triparty*, and, further, between the development of commercial techniques and those of the sciences of calculation. Even if, in the current state of research, the hypotheses are still quite fragile, two remarks deserve to be remembered.

1. Chuquet, like many other authors of commercial arithmetics, was, through his work as an *algorist* and as a teacher of calculation, a professional mathematician, a category of persons which appears to have hardly existed in the West before the fourteenth and fifteenth centuries and, moreover, a mathematician who faced problems of calculation daily. The appearance of such a professional group, linked with the development of the needs of commerce, first of all in Italy, then in France and elsewhere,[72] is for me one of the essential factors in the advancement of arithmetic and algebra.

2. It seems particularly interesting that Nicolas Chuquet separates his exposition of the science of numbers at the beginning of the *Triparty* from its application to commerce, which was transferred to the end of the treatise. The other French and Italian treatises of commercial arithmetic presented the mathematical and commercial aspects in the same chapters; arithmetic and commerce developed reciprocally; the link was a close and almost an organic one. Here the change is radical; the mathematical work exists in its own right and it is applied subsequently to commerce. The period of innovatory commercial arithmetics seems to have ended with the generation of Luca Pacioli and Nicolas Chuquet. From the sixteenth century, the number of commercial arithmetics will increase, but they will not add anything further to the development of mathematics. From the end of the fifteenth century, humanists became interested in arithmetic, but they blended the science of calculation, developed in the treatises intended for merchants, with speculative arithmetic, and they wrote in Latin for a different audience.

---

[71] The question arises of deciding whether this text was directly intended for pupils; the fact that many of the basic principles are supposed to be known makes it rather a 'teacher's book'.
[72] P. Benoit, 'La formation . . .', *op. cit.*, p. 224.

For example, in 1495, Pedro Sanchez Ciruelo, editing Bradwardine's *Arithmetica speculativa*, added a treatise whose contents are quite similar to the commercial arithmetic.

The application of the science of numbers in matters of commerce appears then as a work intended for teaching, embedded in the world of commerce, the inheritor of an extensive tradition of commercial arithmetics but also the last, or one of the last, commercial arithmetics to have been written by a genuine mathematician.

Chuquet's mathematical executor: could Estienne de la Roche have changed the history of algebra?

BARBARA MOSS

1. Introduction

Estienne de la Roche was the author of the first published work on algebra by a Frenchman. In his *Larismethique nouellement composee* (1520),[1] he incorporates, almost verbatim, large sections of Chuquet's manuscript. Aristide Marre, in his edition of the *Triparty* . . . (1880), draws attention to a number of these passages, and concludes:

Without being accused of injustice or exaggeration, one may say that he appropriated the work of Nicolas Chuquet, that he purely and simply copied the *Triparty* in a host of places, that he suppressed certain of the most important passages, especially in algebra, that he curtailed or prolonged others to make his *Arismethique* vastly inferior to the *Triparty*, and, finally, that if Nicolas Chuquet and his work have remained in obscurity for four hundred years, it is above all to him that the primary cause must be attributed.[2]

The only point of this diatribe with which other historians have taken issue is the last clause; Chuquet, according to two popular writers (J. W. N. Sullivan and Howard Eves),[3] was too far ahead of his time.

I shall argue that the charge of plagiarism against Estienne de la Roche is largely an anachronism, and that the difference in emphasis may be

---

[1] Published in Lyon by Constantin Fradin in 1520. A second edition was published in 1538, also in Lyon, by the brothers Huguetan.
[2] A. Marre, 'Notice sur Nicolas Chuquet et son *Triparty en la science des nombres*', *Bullettino di bibliografia e di storia delle scienze matematiche e fisiche 13* (1880), p. 580.
[3] J. W. N. Sullivan, *The History of mathematics in Europe*, Oxford University Press, London, 1925, p. 35: 'He was in advance of his time, particularly in France, which lagged behind Italy in the art of calculation, so that his influence was slight.' Howard Eves, *An Introduction to the history of mathematics*, Rinehart, New York, 1953, p. 214: 'His work was too advanced, for the time, to exert much influence on his contemporaries.' This is a simplification of Moritz Cantor's view: see M. Cantor, *Vorlesungen über die Geschichte der Mathematik* II (second edition, 1900), Teubner, Leipzig, pp. 347–374. Jean Itard, in his article on Chuquet in the *Dictionary of scientific biography*, expresses a kinder variant of Marre's view: 'If Estienne de la Roche had been more insensitive—especially if he had plagiarized the "règle des premiers" and its applications—mathematics would perhaps be grateful to him for his larcenies. He was unfortunately too timid . . .'

explained by the readers he had in mind and by the climate of mathematical opinion, while the similarities are close enough to indicate how far his potential readership would have been receptive to a more faithful presentation of Chuquet's ideas. It is thus possible to estimate how far such a presentation could have affected the history of algebra, and hence to delineate Chuquet's place in that history.

2. De la Roche's sources

Estienne de la Roche can almost certainly be identified[4] as a near neighbour of Chuquet in Lyon, who was still below the age of majority when the *Triparty* was written. He may have been a pupil of Chuquet's, perhaps even a pupil-teacher giving instruction in simple arithmetic while working through the algebraic material of the *Triparty*. His handwriting has been identified in the margins of the only surviving copy of that manuscript, where he jotted down not only his comments on the text but also some additional problems which occur in his book.

In 1484, when Chuquet's manuscript was completed, a mere handful of titles had been published in Lyon; by the middle of the following century the town had become the third most important centre for the book trade in Europe after Venice and Paris, and particularly famous for scientific and technical books. A substantial arithmetic by the Spaniard Juan Ortega was translated into French and published in Lyon in 1515, and this may have suggested to de la Roche that the time was ripe for publication of his own work, which appeared five years later.

Before the spread of printing, academic knowledge had been disseminated through the copying of manuscripts, and Chuquet, like many of his contemporaries, must have written down for future reference a large number of examples from the work of others, with or without a note of their source. He cites very few works, and those, almost always, in order to attack them. Pacioli, who has frequently been accused of plagiarism, is more generous, naming several sources, the most important of whom was Fibonacci, and his *Summa* includes passages taken from the *Liber abaci*, as well as a whole section copied without acknowledgement from a manual for merchants by Chiarini.

De la Roche's use of citations and sources is similar to that in a number of printed arithmetics of that period. Following the usual commendation of mathematics for its 'great utility and necessity', he continues:

---

[4] H. L'Huillier, 'Éléments nouveaux pour la biographie de Nicolas Chuquet', *Revue d'Histoire des Sciences* 19 (1976), pp. 47–350. See also his *Nicolas Chuquet: La Géométrie*, Vrin, Paris, 1979 pp. 78–9 for a discussion of the handwriting of the marginal annotations on Chuquet's manuscript.

I have collected and amassed the flowers of several masters expert in this art, such as Master Nicolas Chuquet, Parisian, Philippe Frescobaldi, Florentine, and Brother Luke of Burgo Sancti Sepulchri of the Order of Friars Minor, with some small addition of what I have been able to invent and test out in my time in its practice.[5]

Brother Luke is, of course, Luca Pacioli. Frescobaldi, as L'Huillier has discovered, was a banker in Lyon; although no manuscript of his is known, he was probably the source of some of the specifically French commercial material that is not in Chuquet, and he may also have been able to provide de la Roche with access to some Italian manuscripts. (Professor Franci has pointed out that de la Roche's algebraic notation bears a closer affinity to that of the Florentine school, as represented by Ghaligai, than to Pacioli.)

At worst, then, de la Roche cobbled together parts of the works of three authors, which were almost certainly inaccessible to the average French merchant and which were not protected by author's copyright. (Many early books, like de la Roche's own *Arismethique*, carried only a publisher's copyright, or privilege, extending over a short period—two years in his case.) The second edition, in 1538, was further improved by the inclusion of a set of arithmetical tables compiled by the publisher, Gilles Huguetan. Apart from ready reckoners and tables for calculating the cost of a regular payment over various time periods, these included special tables for printers and booksellers.

2.1. *Chuquet, Pacioli, and de la Roche—distinctive features*

The contrast between de la Roche's two known sources has been examined by several authors. One modern account, due to Juschkewitsch (1961, 1964) and based on the *Triparty* alone, reads as follows:

Although Chuquet's work is not as comprehensive as the *Summa* of Pacioli, although it is not as encyclopedic and presents significantly fewer examples, yet in terms of the ideas it contains it is essentially more abundant. Pacioli was merely the teacher who had thoroughly mastered all the mathematical knowledge to which he had access. Chuquet was an original thinker and creator of new generalized concepts which have driven algebra forward.[6]

This is an over-simplification; Chuquet's entire manuscript, not just the algebra, should be set against the *Summa*, and it is also relevant to consider the different circumstances in which the two authors were writing. Three aspects will be examined here: the scope of the two texts, the pedagogic skill with which they are presented, and the originality of the Frenchman's work.

---

[5] De la Roche (1520), first (unnumbered) page: '*ay collige et amasse la fleur de plusieurs maistres expertz en cest art: comme de maistre nicolas chuquet parisien: de philippe friscobaldi florentin: et de frere luques de burgo sancti sepulchri de lordre des freres mineurs avec ques quelque petite addicion de ce que iay peu invente et experimenter en mon temps en la pratique*'.

[6] A.P. Juschkewitsch, *Geschichte der Mathematik im Mittelalter*, Teubner, Leipzig, 1964, p. 433. This book was originally published in Russian in 1961.

Chuquet's manuscript is indeed less comprehensive than Pacioli's; but it does include two sections neglected by Juschkewitsch because they had not then been printed: the *Commercial arithmetic* and the *Geometry*. De la Roche devoted roughly two-thirds of his book to the former topic, and incorporated a number of examples not only from the corresponding section of Chuquet's manuscript but also from the more algebraic *Problems*. His material on gold and silver comes direct from Chuquet. In general, he goes into more detail about the settings of his problems: a merchant buys 'wool from Perpignan' instead of just 'cloth'. Pacioli's influence is most apparent in the more sophisticated trading practices, such as various kinds of barter, some of which were thinly disguised forms of lending money at interest, or usury. However, de la Roche does not include Pacioli's famous section on book-keeping; perhaps the practice had not yet reached the French merchants in Lyon.

Whereas Chuquet had devoted much space to geometrical constructions and to geometric problems solved by algebraic methods, de la Roche includes very little on geometry. His short chapter on the subject could have been derived from any one of a number of sources, and he does not even present Chuquet's discussions of the rudiments of such practical subjects as surveying and gauging.

It is not surprising that Pacioli's work is a better textbook than Chuquet's. Pacioli had been teaching mathematics since his youth, first in a merchant's family, later in universities and through public lectures, and he had written a manuscript for his young pupils more than twenty years before the *Summa* appeared. Moreover, he was writing for publication, whereas the *Triparty* manuscript reads more like a teacher's handbook, with no attempt to intersperse theory and applications. De la Roche, also, was writing for publication, and his book is more attractively laid out than either of his sources.

The last distinction drawn by Juschkewitsch is that of originality of thought and notation. Now, this may be partly a consequence of Chuquet's position, living in a town with no tradition of university education and writing mainly, if not entirely, for his own benefit, in a language with no algebraic vocabulary. All these factors allowed him greater freedom to experiment than Pacioli; he had to invent the language, so he came to invent the terminology and notation. (In the same way, German algebras and arithmetics of the early sixteenth century were more inventive in notation than their Italian counterparts, establishing the symbols + and − instead of the constructions $\bar{p}$ and $\bar{m}$. Even England, a latecomer indeed, provided the multiplication ($\times$) and equals (=) signs, and Scotland, perhaps, the decimal point.) This is not to belittle Chuquet's achievement, but it may help to explain why de la Roche, writing for a community in which the commercial supremacy of the Italians was beyond dispute, reverted to more traditional names for algebraic processes, and to notations similar to those of the Florentine school.

3. De la Roche's treatment of Chuquet

3.1. Arithmetic

As far as the first book of the *Triparty* is concerned, de la Roche is reasonably faithful to his teacher. He does include an extra chapter, on the connotations of the numbers from 1 to 12, which he took from Pacioli, and Pacioli from St Augustine's *Civitas Dei*. He also prefers some of the more conventional names, like 'rule of false position' rather than 'rule of one position', and disagrees with Chuquet's assessment that the rule of apposition and remotion for solving indeterminate equations in integers 'is a science of little recommendation'.[7] However, he includes two of the distinctive contributions of the *Triparty*: nomenclature in terms of the powers of 10^6 up to the nonillion, and the rule of intermediate terms, or 'rule of mediation between the greater and the less'. The first, so far as I know, was not taken up by any of his readers, though Buteon applies the second in his *Logistica*.[8]

Chuquet devotes much more space than de la Roche to the calculus of radicals and surds. In particular, de la Roche says very little about higher-order roots, which he regards as 'not of great utility',[9] and he omits the table of the first ten powers of numbers from 1 to 10. His notation for radicals is a combination of Chuquet's notation and the one for second, third and fourth roots which Chuquet had criticized as too limited; in fact, he uses the latter rather than the former wherever possible in his subsequent working.

Worse, although he reproduces Chuquet's definition of bound roots and states that the terms bound together should be underlined, he does not show them as such in either edition of *Larismethique*. Thus the invention of bracketing was squandered by de la Roche, and disappeared until the middle of the sixteenth century when a similar notation was developed by Bombelli.[10]

3.2. Algebra

De la Roche's algebra contains several topics that are not found in the *Triparty*, though some of these are developed in Chuquet's solutions to the *Problems*. As examples of the latter, he has chapters on algebraic fractions,

---

[7] '*apposicion et remocion est science de petite recommandacion*' (Chuquet, MS. **BN fonds français 1346** f.4ʳ; see Flegg, Hay and Moss, eds., *Nicolas Chuquet, Renaissance mathematician*, Reidel, Dordrecht, 1985, p. 90).

[8] J. Buteo or Buteon (Jean Borrel), *Logistica, quae et arithmetica vulgo dicitur*, 1559.

[9] '*Les racines quintes sixiesmes septiesmes etc ne sont pas de grande utilite.*' (De la Roche 1520 f.34ᵛ)

[10] For Bombelli's ' bracketing' notation, see F. Cajori, *A history of mathematical notations*, volume I, Open Court, 1928, p. 128 (figure 52).

¶ Emores de .20. Je veulx faire deux parties telles que
multipliees chascune en soy et puis soustraire la
mineur de la maieur la reste soit .50. Pour ce faire
je pose que la moindre partie soit .1.¹ Ainsi la plus grande
sera .20.m̃.1.¹ Or, multiplions chascune en soy si auros

1.² pour la premiere multiplication et .400.m̃.40.¹.p̃.1.² pour la
seconde. Opés soustrais .1. de .400.m̃.40.¹.p̃.1.² Restent
400.m̃.40.¹ egaulx a 40 (¶ Donne maniere) 40. a l'une
et a l'aultre parties si auras .40.p̃.40.¹ Ounne part et
400. d'aultre. Puis oste .40. de chascune partie et auras
360. pour nombre a partir et .40. pour partiteur. Partiz
donc 360. par 40. si auras .8.½. pour la plus grande
et moindre partie. Et par consequent .11.¼. pour l'aulte

¶ Qui autrement par Regle speale, partiz .40. par .20. vient
a la part .2.⅟₂. Levez de .20. Restent 17.½. dont la
moitié qui est .8.¾. est l'une des parties. Et par ainsi
l'aultre sera .11.¼. comme dessus.

¶ Plus partez .20. en deux parties telles q̃ multipliee chascūe en soy. Et puis soustrai
re la mineur de la maieur. La reste soit. 50. Responce. Pose que la maieur partie soit. 1.
p̃. Ainsi la mineur sera. 20. m̃. 1. p̃. multiplie chascune partie en soy/et auras. 1. ce̊. pour
la maieur partie/7. 400. m̃. 40. p̃. p̃. 1. ce̊. pour la mineur. Ores de. 1. ce̊. soustraiz. 400.
m̃. 40. p̃. p̃. 1. ce̊. reste. 1. ce̊. m̃. 400. p̃. 40. p̃. m̃. 1. ce̊. Qui abbreuiez sont. 40. p̃. m̃.
400. egaulx a. 50. donnez. 400. dung coste/ʒ daultre. Et auras. 40. p̃. egales a. 450.
Partiz. 450. par. 40. ʒ en vient. 11. ¼. pour la maieur partie. Et par consequét, 8. ¾.
pour la mineur.

1. p̃.————20. m̃. 1. p̃.	1. ce̊. m̃. 400. p̃. 40. p̃. m̃. 1. ce̊.		
1. p̃.————20. m̃. 1. p̃.	40. p̃. m̃. 400.————50		4 5|0
1. ce̊.— 400. m̃. 40. p̃. p̃. 1. ce̊.	400		1 1
	450		4|0

FIGURE 1. A simple problem in algebra, solved by Chuquet (top) and la Roche (bottom)

on the rule of quantity (a mixture of algebra and the rule of false position used for solving equations in more than one unknown), and on equations with more than one solution. He also gives methods for 'proving', or checking, results in algebra analogous to Chuquet's 'proofs' in arithmetic.

The four canons of the rule of first terms (one for solving generalized linear equations, and three for the three acceptable forms of the generalized quadratic, with all coefficients positive) are stated in Chuquet's terms, although de la Roche gives more elementary examples, and fewer which require solution by means of compound roots or roots of high order. However, special cases of cubic equations, with no constant term, are included among the quadratics.

The fourth canon, for solving equations of the form $x^2 + b = ax$, which may have two positive roots, gave rise to a minor controversy. In 1559, Jean Buteon attacked de la Roche's rule, and claimed that it is impossible for an equation to have more than one solution.[11] De la Roche's example to the contrary should be disallowed because he gives 1 as one of the roots, and 1, according to Buteon and a number of his predecessors and contemporaries, should not be considered as a number. De la Roche also gave an example with roots 6 and 4; would Buteon have disallowed this because the roots were computed as $5 + 1$ and $5 - 1$? These examples are more accessible to the beginner than Chuquet's problems on the same canon, of which the first has no real root, the second has a double root, and the third has solutions $5 + \sqrt{21}$ and $5 - \sqrt{21}$. On equations with no real roots, de la Roche remarks:

And thus on every proportion, innumerable impossible questions may be posed, to test out the masters and to know if they are expert in proportions.[12]

Despite the conceptual regress in his notation, de la Roche's way of separating out the symbolic working from the rhetorical description of the method makes his book easier to follow than Chuquet's manuscript. Figure 1 compares the solutions given by Chuquet and de la Roche to the following problem:

Divide 20 into two parts such that when each is multiplied by itself and then the lesser subtracted from the greater, the remainder is 50.

So much for the strengths of de la Roche's algebra. The weaknesses are essentially those of conservatism, though his work, like Chuquet's, is not free from careless errors. As in the case of radicals, he presents the concepts of algebra both in the terminology and notation of the *Triparty* and in a more traditional, restricted, and qualitative system involving symbols not

---

[11] Buteo, *Logistica, supra*.
[12] De la Roche (1520) f.70ʳ: '*Et ainsi sus chascune proporcion se pourroyent proposer innumerables questions impossibles pour tater les maistres et pour scavoir silz sont expertz aux proporcions.*'

numbers for powers of the unknown, and he prefers to use the latter. This makes nonsense, for instance, of the table of powers of 2 in which Chuquet shows his understanding of the index laws; below the fifth power, numerical indices do not appear.[13]

However, the index notation *is* presented. It attracted the attention of Michel Chasles (see below, § 5) in the nineteenth century. Was it observed by any of de la Roche's contemporaries? Was Bombelli's index notation based on Chuquet's? It is not possible to give a definite answer to this question, only to trace the known readers of de la Roche's *Larismethique* and the use, if any, that they made of its peculiar features.

4. De la Roche's readers

Three French mathematicians of the sixteenth century cite de la Roche: Jean Buteon in his *Logistica* of 1559, Fulconis in *Opera nove d'arismetica* (1562), and Gosselin in *De arte magna* (1571). Of these, Buteon's citation, discussed in the previous section, is the most interesting, and is picked up by Bonasoni, who speaks out in vindication of de la Roche. This also ensured that the name (in a distorted form) was handed down to posterity in Wallis's *History of algebra* (1695). Is it possible that Bonasoni knew de la Roche only through Buteon's citation; if not, then at least one copy of *Larismethique nouellement composee* reached Italy. Moritz Cantor, however, remarks:

At that time, an Italian merchant would have simply regarded as ludicrous (the suggestion) that he should learn the principles of reckoning or the doctrine of equations from a German or a Frenchman.[14]

There are further references to de la Roche in Leurechon's work on mathematical recreations (1624; sometimes attributed to van Etten), in a Dutch work on book-keeping which I have not seen, and in the bibliographies of early printed books on mathematics compiled by Panzer and by Heilbronner. The latter appends the comment: *Opus esset optimum si demonstrationes haberet* ('The work would be very good if it had proofs').

Humphrey Baker does not refer to de la Roche in his *Wellspring of science*, but he does introduce the rule of three by exactly the same example as de la Roche and Chuquet ('If 8 are worth 12, what will 14 be worth?'), and the example is too abstract to be entirely natural.

Cardano's description of his rule of method, for finding special algebraic rules for a class of problems by consideration of numerical examples, is

---

[13] *ibid.* f.43^v.
[14] M. Cantor, *Vorlesungen über die Geschichte der Mathematik* II (second edition, 1900), Teubner, Leipzig, p. 347.

similar to Chuquet's.[15] However, I have not found the passage in de la Roche. Cardano himself claims that Pacioli used this technique, but here again I have been unable to find the reference. This early precursor (in spirit, at least) of Polyà's *How to solve it?* deserves to be tracked down more closely.

Finally, Bombelli, in a manuscript written in the middle of the sixteenth century, exhibits similar notations both for indices and for the bracketing of the terms in a compound root; but he does not use Chuquet's terminology for powers, which he might be expected to have adopted with enthusiasm had he known of it, nor the Frenchman's notation for roots of higher order than 2. The balance of probability, then, favours an independent development of notation on Bombelli's part.[16]

5. Michel Chasles' account of the history of algebra

In 1841, the French geometer Michel Chasles, who had already written a substantial history of geometry, presented three papers on the history of algebra to the Paris *Académie des Sciences*.[17] His main theme was the achievement of Viète in establishing a literal, or symbolical, algebra, that is, a calculus based on symbols, not merely as abbreviations for the names of variables, but as entities on which arithmetical laws could be defined and arithmetical operations performed.

Chasles saw no anticipation of Viète's ideas among the Italians; but he instanced the German Stifel and the Frenchmen Peletier and Buteon, because they used letters for unknowns and a crude form of index notation. The development of an adequate notation for exponents had hitherto been accredited to Descartes; but Chasles claimed that such a notation is already present in de la Roche's *Larismethique nouellement composee*.

Unfortunately, Chasles got the details wrong, believing that $12.^2$, $12.^3$, etc. represented powers of 12, whereas in fact they stand for $12x^2$, $12x^3$, and so on. Chasles also failed to notice the terminology for first, second, third and

---

[15] Girolamo Cardano, *Ars magna*, 1545, chapter XXIX. To quote from the translation by T. Richard Witmer (*The great art or the rules of algebra*, p. 180: 'This, then, is the rule [of method]: Solve any given problem by any means you can . . . Then lay aside the unknown and the other rules and use those operations which you can best use, and you will have the rule for the method of "solving" any similar problem.' Compare Chuquet, ms f.195^r: 'The style and manner of seeking and making such special rules is that one should carry out . . . two or three problems of a similar nature with various numbers. And then . . . reflect on the known numbers, how they might have been found had they been unknown . . . until one has found a means by which one might arrive at the numbers . . .' This passage is translated in full in Flegg *et al, op. cit.*, p. 224.

[16] Cajori, *op. cit.*.

[17] M. Chasles, 'Note sur la nature des opérations algébriques . . . ', *Comptes rendus de l'Académie des Sciences*, 12 (1841), pp. 741–756; 'Sur l'époque où l'algèbre a été introduite en Europe', *ibid.*, 13 (1841), pp. 497–524; 'Sur les expressions *res* et *census*', *ibid.*, pp. 602–628. The main reference to de la Roche is on p. 752.

higher powers, in which he could have observed a significant improvement on the old terminology derived from Hindu and Arabic algebra:

> This most imperfect system was . . . still employed in all their [European Christians'] works of the sixteenth century. The most celebrated analysts of that epoch, Cardan, Tartaglia, and Bombelli, only knew this imperfect nomenclature, which halted the development of algebra.[18]

If de la Roche had been more faithful to Chuquet's notation and terminology and had used it consistently, and especially if this had given rise to some controversy in the literature, then Bombelli's work, at least, might have been enhanced. In this sense, and in this sense only, can de la Roche be held responsible for Chuquet's ideas remaining in obscurity.

6. Chuquet's place in the history of algebra

In spite of his valuable insights into the generality of algebra, Chuquet's work suffers from many of the defects of his age. He gives few formal proofs, and admits to formulating rules by crude induction, leading to several errors.

The major influence on the development of algebra during the late sixteenth and seventeenth centuries was the rediscovery of the work of Pappus and Diophantus. Bombelli, Viète, Fermat, and Descartes were all inspired by this work, and Viète in particular was concerned to restore the standards of rigour which had been lost, and to reconstruct mathematics on the Greek model.

But if there had been no algebra in the preceding hundred years, this reconstruction would not have occurred. It was the shortcomings of their immediate predecessors that sent them back to the Greeks; but it was the achievement of these same predecessors that algebra was a live subject deemed worthy of reconstruction. Although Viète did not adopt Bombelli's notation, the quest for such a notation, finally established by Descartes, had been under way since the *Triparty*, and Chuquet already saw the potential for exploiting a generalized index notation and terminology to free algebra from the straitjacket of geometrical interpretations.

[18] *ibid.*, *13* p. 604.

How algebra came to France

WARREN VAN EGMOND

ALGEBRA as we know it began in France around the beginning of the seventeeth century. Although its roots are very old, reaching back to the problem-solving techniques of ancient Mesopotamia and the analytical methods of the Hellenistic world, it was the conceptual transformations wrought by François Viète and his immediate followers, together with the new symbolism invented by Descartes, that established algebra in the form we know today. Since that time algebra has been the fountainhead of European mathematics, the source from which the most important branches of modern mathematics have sprung. Analytic geometry, the calculus, and all of the varied disciplines of modern analysis flowed directly from the concepts, methods, and forms of algebra that were first set out in France around the turn of the seventeenth century.

The problem that historians of mathematics face is to explain why this event or rather this series of events took place at just this time and place. Since algebra is nearly as old as recorded history itself, we would like to know what combination of circumstances brought algebra to France and provided just the right set of factors that produced such a revolutionary change in its character and form. What were the conditions and events that preceded and presumably spawned this extraordinary transformation?

We already have a fairly clear understanding of how algebra came to Europe as a whole. As its name indicates, the Europeans took their knowledge of algebra from Islam, where it had first been described by the great Arab mathematician and astronomer Muhammed ibn Musa al-Khwarizmi in the ninth century AD[1] and then further developed by many able Islamic mathematicians in the following centuries, culminating in the solution of the cubic equation by the famous poet and mathematician Omar Khayyam in the eleventh and twelfth centuries.[2] The basic concepts and methods of algebra passed to the West in the twelfth century as part of the great movement for

---

[1] *Kitāb fī hisāb wa'l muqābala*. An Arabic text with English translation was published by Frederic Rosen, *The algebra of Mohammed ben Musa* (London, 1831).

[2] See the Arabic text and French translation by Roshdi Rashed and Ahmad Djebbar, *L'Oeuvre algébrique d'al-Khayyām* (University of Aleppo, 1981), which supersedes the older French translation by F. Woepcke, *L'algèbre d'Omar Alkhayyâmî* (Paris, 1851). There are English translations by Daoud S. Kasir, *The algebra of Omar Khayyam* (New York: Columbia University, 1931, reprinted New York: AMS Press, 1972), and by H. J. J. Winter and W. ᶜArafat, 'The Algebra of ᶜUmar Khayyām', *Journal of the Royal Asiatic Society of Bengal*, 16 (1950), pp. 27–77.

he translation of Arabic science and philosophy into Latin. Discussions of algebra are found, for example, in such works as Savasorda's *Liber embadorum* (translated by Plato of Tivoli in 1145), Abraham ibn Ezra's *Liber augmenti et diminutionis*, Abû Bekr's *Liber mensurationum*, and John of Spain's *Liber algorismi de practica arismetica*.[3] But three works stand out for the importance they had in transmitting Arabic algebra to the West—the two translations of al-Khwarizmi's text made by Gerard of Cremona and Robert of Chester in Spain in the twelfth century,[4] and the *Liber abbaci* of Leonardo Pisano, which was completed in Italy in 1202.[5] The latter was not a translation of any single book but rather a massive compendium of Arabic mathematics based on the extensive knowledge of this field that Leonardo had acquired from his studies under Arab teachers as a boy in North Africa, where his father was the manager of the Pisan customs house in Bugia, and during his later travels in the Arab world.[6] The last part of the *Liber abbaci*, part 3 of Book 15, contains a discussion of algebra drawn from the works of al-Khwarizmi, abu-Kamil, and al-Karaji.[7]

Following this initial period of translation and transmission, which was limited to the fundamental works of the early Arab algebraists and did not include the later, more sophisticated work of Omar Khayyam or Sharif al-Din al-Tusi, for example, the centre for the study of algebra in Europe remained in Italy, contained within the larger tradition of abbacus mathematics. This tradition is logically a continuation of the work of Leonardo Pisano but historically begins at the end of the thirteenth century with the appearance of the first abbacus schools and the *maestri d'abbaco* who taught in them, which were devoted to teaching the principles and methods of elementary arithmetic to the sons of the growing Italian merchant and com-

---

[3] M. Curtze, 'Der "Liber Embadorum" des Abraham bar Chijja Savasorda in der Übersetzung des Plato von Tivoli', *Abhandlungen zur Geschichte der mathematischen Wissenschaften*, 12 (1902), pp. 1–183. See also Martin Levey, 'Abraham Savasorda and his algorism', *Osiris*, 11 (1954), pp. 50–63; and 'The encyclopedia of Abraham Savasorda: a departure in mathematical methodology', *Isis*, 43 (1952), pp. 257–264. The text of Abraham ibn Ezra was published by Guillaume Libri, *Histoire des sciences mathématiques en Italie*, vol. I (Paris, 1838), pp. 304–371. H.L.L. Busard, 'L'algèbre au Moyen Age: le "liber mensurationum" d'Abû Bekr', *Journal des savants* (1968), pp. 65–124. John of Spain's text was published by Baldassarre Boncompagni in his *Trattati d'aritmetica* (Roma, 1857).
[4] The text of Gerard of Cremona was published by Libri, *op. cit.* pp. 253–297. Robert of Chester's version was published by Louis C. Karpinski in *Contributions to the history of science* (University of Michigan Studies, Humanistic Series, 11, Ann Arbor, 1930).
[5] Text published by Baldassarre Boncompagni in *Scritti di Leonardo Pisano*, vol. I (Roma, 1857).
[6] See Richard E. Grimm, 'The autobiography of Leonardo Pisano', *Fibonacci quarterly*, 11 (1973), pp. 99–104.
[7] *Op. cit.* (note 5), pp. 406–459. This section was also published by Libri, *op. cit.* (note 3), pp. 253–297. The precise identification of all the sources used by Leonardo remains a subject for future research.

Provençal *Art del algorisme* which has recently been studied by Dr Sesiano.[31] It was written in Pamiers, a town on the main road between Toulouse and Barcelona, about 1446, although its original date of composition appears to revert to the previous decade. Its organization and content are typical of an abbacus book. It is divided into three parts—an algorism of integers, an algorism of fractions, and a presentation of the three principal rules of the abbacus (algebra is not included), all illustrated with numerous problems of a commercial nature. A second manuscript, the *L'art d'arismetique*,[32] also dates from this period, but it has not yet been possible to identify the locale of its composition.

But we must wait yet another quarter century, until 1475, before the French abbacus tradition reaches its peak of activity. Within the ten-year period from 1475 to 1485 fully six important manuscripts were written: the *Traicte en arismeticque* of Jehan Adam, which was written in Paris in 1475;[33] the important compilation of five works made by Matthew Prehoude in Lyons in 1476;[34] the work that begins 'Toute bonne science . . . ', which may have been written by Jean Fusoris in Normandy in 1477;[35] Nicolas Chuquet's *Triparty en la science des nombres*, whose quinquecentenary provides the occasion for this conference;[36] and the *Traicte de algorisme* of Jehan Certain, which was written in Bilbao in 1485.[37] The anonymous *Cadran aux marchans*, which was heavily based on Certain's work,[38] may also be added to this list, as well as the *Compendion de lo abaco* of Francesco Pellos which, although it was published in Turin in 1492, was originally composed in Nice in a Provençal dialect in the 1460's. All of these works share enough similarity of content and organization to demonstrate that they belong to a single tradition of practical mathematics that is distinctively French in character.

One of the key figures in the formation of this French abbacus tradition appears to have been Bartholomieu de Romanis or Roumanis, an otherwise

---

[31] Paris, Bibliothèque Nationale, ms. français nouv. acq. 4140, 16^r–117^r. See Jacques Sesiano, 'Une arithmétique médiévale en langue provençale', *Centaurus*, 27 (1984), pp. 26–75.
[32] Paris, Bibliothèque Nationale, ms. Français 2050, 151 cc.
[33] Paris, Bibliothèque Ste-Geneviève, Ms. 3143. See Lynn Thorndike, 'The arithmetic of Jehan Adam', *American mathematical monthly*, 33 (1926), pp. 24–28, and his *Science and thought in the fifteenth century* (New York, 1963), pp. 151–160 and 302–307.
[34] Cesena, Biblioteca Malatestiana, ms. S. XXVI. 6.
[35] Paris, Bibliothèque Nationale, ms. Français 1339, 1^r–114^v.
[36] Paris, Bibliothèque Nationale, Ms. Français 1346. Parts published by Aristide Marre, 'Notice sur Nicolas Chuquet et son Triparty en la science des nombres', *Bullettino di bibliografia e di storia delle scienze matematiche e fisiche*, 13 (1880), pp. 555–659, 693–814; 14 (1881), pp. 417–460; and by Hervé l'Huillier (ed.), *Nicolas Chuquet: la géométrie. Première géométrie algébrique en langue française (1484)* (Paris: Vrin, 1979).
[37] Paris, Bibliothèque de l'Arsenal, Ms. 2904. See Paul Benoit, 'La formation mathématique des marchands français à la fin du Moyen Age: l'exemple du Kadran aux marchans (1485)', in *Les entrées dans la vie. Initiations et apprentissages* (XIIe Congrès de la Société des historiens médiévistes de l'enseignement supérieur public—Nancy 1981) (Presses Universitaires de Nancy, 1981), pp. 209–224.
[38] Paris, Bibliothèque Nationale, ms. Français nouv. acq. 10259, 5^r–98^v.

unknown Dominican preacher from Valence, whose work is cited by both Jehan Adam and Nicolas Chuquet.[39] We are fortunate enough to possess a copy of one of his works in the important compilation made by Matthew Prehoude in Lyons in 1476, in which he describes himself as a 'predicateur en avoye'.[40] Prehoude's compilation also includes a work by an unnamed 'master of Carcassonne',[41] further indicating the geographic extent of the movement.

We can only speculate on the reasons for this sudden burst of mathematical activity in France in the last quarter of the fifteenth century, or the sources from which it sprang. In view of the fact that Italy was at this time a flourishing centre for practical mathematics, as it had already been for the better part of two centuries, it is natural to suppose that the chief inspiration came from this direction. Indeed, the geographical focus of the movement in the South of France seems to add *prima facie* support to this hypothesis. Southern France, Provence, and Lyon in particular were those areas of France most closely linked to Italy through trade routes, and there was even a permanent Italian community living in Lyon.[42] However, it is difficult to find any firm evidence that would convincingly demonstrate an Italian origin for the French abbacus manuscripts. There are no direct references to any Italian works or known Italian authors in any of the extant manuscripts, and the few linguistic parallels in terminology or phrasing do not provide convincing evidence of literal dependence on Italian sources. Barring a lengthy and detailed comparison of all the extant French and Italian manuscripts, it would be difficult to prove the existence of this connection beyond all doubt.[43]

In fact, there is one feature of the French texts of this period that clearly sets them apart from the Italian abbacus manuscripts. In addition to the three basic methods of problem solving, the rule of three and the rules of single and double false position, almost all of the French abbaci beginning with the Provençal *Art del algorisme* add a fourth method that they call the 'regle d'apposition et de remotion',[44] which has no counterpart in any Italian

---

[39] Ste-Geneviève 3143, ff. 3^r, 13^r. français 1346, f. 186^v, pub. in Marre, *op. cit.* (ref to note 36), Vol. 14, pp. 416 note 1 and 442.

[40] Cesena S. XXVI. 6, ff. 149^r–268^v.

[41] *ibid.*, ff. 269^r–290^v.

[42] It has been estimated that by the sixteenth century, when reliable figures first became available, Italians represented 5 per cent of the population of Lyons and controlled 28 per cent of the taxable wealth. See Richard-Félix Gascon, 'Les Italiens dans la renaissance économique lyonnaise au XVIe siècle', *Revue des etudes Italiennes*, (N.S.) 5 (1958), pp. 167–181, especially p. 172. René Jullian,'Lyon et l'Italie au Moyen-Age (histoire et art)', *ibid.*, pp. 133–146, remarks on p. 145: 'C'est ainsi qu'au cours de la seconde moitié du XVe siècle une puissant colonie italienne se constitue à Lyon, faite essentiellement de marchands et de banquiers'.

[43] It might be noted that the copying of manuscripts in places like Carcassonne, Pamiers, and Bilbao points as much to a possible Spanish origin as to an Italian one.

[44] See Sesiano, *op. cit.* (note 31) pp. 61–63. Chuquet's description of this method is in Marre, *op. cit.* (ref to note 36), Vol. 13, pp. 652-3.

text of the period. It is in fact a rather ingenious method for finding an integral solution to an indeterminate problem of two equations with three unknowns, which is illustrated by a few problems that are shared among most of the manuscripts. The phrase 'apposition et remotion' is tantalizingly similiar to the phrase 'oppositio et restauratio' that was sometimes used as a name for algebra in medieval Latin texts (it is a literal translation of the Arabic *al-jabr wa'l muqabala*),[45] but the method is not algebraic in any way. As with the French abbaci themselves, it is impossible to say where it came from.

One possible source for the French abbacus tradition that seems to be worth considering is the native Hebrew tradition of astronomy and mathematics that flourished in the South of France in the fourteenth and fifteenth centuries. Since the beginning of the twelfth century a long string of scholars living in the Jewish communities of Spain and southern France had devoted themselves to the study of astronomy and mathematics, motivated primarily by the sophisticated astronomical computations that are needed in order to calculate the Jewish religious calendar.[46] Many original and important results arose from their work. Nearly half a century ago George Sarton pointed out how strongly Chuquet's use of fractions and his exponential notation resembled the formulations developed by Immanuel ben Jacob Bonfils of Tarascon a century earlier.[47] But again it is difficult to demonstrate that there were any definite connections between Jewish and Christian writers, and there are a number of considerations that make it seem unlikely. For one thing, the Jewish tradition was almost exclusively devoted to astronomy, and shares none of the problems and techniques that characterize the abbacus manuscripts. For another thing, the Jewish texts were all written in Hebrew, which few Christians of that day could read, and the paths by which these Hebrew writings might have reached the abbacists remain unattested. Again, occasional isolated similarities provide no demonstration of definite connections.[48]

The problem of sources faces us again when we look at Chuquet's algebra. As I have said, Chuquet's *Triparty* was the only one of the French abbaci to include a section on algebra and it is also to our knowledge the only algebra composed in France in the fifteenth century. This seems to rule out any indigenous sources. Chuquet calls algebra the 'rigle des premiers',[49] which

---

[45] 'Incipit liber restaurationes et oppositionis numeri . . . ' in Robert of Chester, *op. cit.* (ref to note 4) p. 66.
[46] Moritz Steinschneider, *Mathematik bei den Juden*, originally printed in *Bibliotheca mathematica*, 1893–1901, reprinted by Georg Olms, Hildesheim, 1964.
[47] See George Sarton and Solomon Gandz, 'The invention of the decimal fractions and the application of the exponential calculus by Immanuel Bonfils of Tarascon (c. 1350)', *Isis*, 35 (1936), pp. 16–45.
[48] Gandz himself points out that Bonfils' work had no influence on his contemporaries or successors. *Ibid.*, p. 31.
[49] Marre, *op. cit.* (ref to note 36), Vol. 13, p. 747.

appears to be a direct translation of the standard Italian name 'regola della cosa', where Chuquet has substituted the word 'premier' for 'cosa', as he also substituted 'champs' for 'censo', the standard name of the second power of the unknown, x^2. But beyond this superficial resemblance, Chuquet's approach to algebra is completely different from that found in the Italian algebras of the period. For one thing, Chuquet used a very advanced exponential notation for the powers of the unknown in which he wrote the power above and to the right of the coefficient, just as we do, although he omitted any symbol for the unknown itself. The Italians simply abbreviated their names for the unknown, writing *co.*, *ce.*, *cu.*, etc. Also, the Italians all followed al-Khwarizmi in dividing the solution of equations into a series of special cases formed from all possible permutations of the component terms—*numero*, *cosa*, *censo*, *cubo*, etc.—with negative and zero terms excluded. Each case then had its own rule or solution procedure which was given in the form of an algorithm. Thus there was one possible linear equation, five quadratic equations, thirteen cubic equations, and so on. Chuquet shows considerably greater economy and insight by instead setting up four general types or 'canons' of equations which are based on the arithmetical difference between the powers of their unknown terms.[50] In modern symbolism, Chuquet's four canons are (I omit the coefficients):

1) $x^n = x^n$
2) $x^n + x^{n+m} = x^{n+2m}$
3) $x^{n+m} + x^{n+2m} = x^n$
4) $x^{n+m} = x^{n+2m} + x^n$

Thus for Chuquet, $x + x^2 = x^3$ and $x^2 + x^3 = x^4$ belong to the same class and can be solved by the same rule, whereas in the Italian algebras these equations were listed separately and had their own rules, as did $x^3 + x^4 = x^5$, $x^4 + x^5 = x^6$, and so on.

The only Italian algebra that adopts a similar approach is the *L'agibra* of the Biblioteca Estense di Modena,[51] but this differs from Chuquet's by specifiying numeric differences of one, two, three, and four between the exponents and setting up eighteen specific classes of equations, whereas Chuquet's canons are general for all cases where the exponents of the three terms are in arithmetic proportion. The Modena manuscript was written at about the same time as the *Triparty*, but a comparison of the two works has revealed no signs of borrowing on either side.[52] The Modena algebra was later used by Dionigi Gori, a Sienese abbacist of the sixteenth century, as is

---

[50] *Ibid.*, pp. 748-9.

[51] Ms. Ital. 578. See the description in *Practical mathematics . . . , op. cit.* (ref to note 10) pp. 171-2.

[52] See my forthcoming edition and study of this text in the Quaderni del Centro Studi della Matematica Medioevale, Università di Siena.

revealed by an almost literal rendition of the same cases and problems in his *Libro d'albaco* of 1544,[53] but there is no other indication that this text was known elsewhere in Italy. Similarly, Chuquet's algebra had only one follower in France, the *L'arismetique* of his probable student and successor, Estienne de la Roche, which was published in Lyons in 1520 and 1538. De la Roche changed the name of algebra back to 'regle de la chose', thus returning to the original Italian phrase, and he preferred the Italian symbolism to Chuquet's highly original and prescient exponential notation, but he did retain Chuquet's four canons and most of his principles.[54]

The tradition of French practical arithmetic that appeared so suddenly and so mysteriously in the last quarter of the fifteenth century, springing up as if from nowhere, unfortunately faded and disappeared as quickly as it had come, leaving few traces of its brief existence. No further treatises were written after about 1485, except for de la Roche's published work, and all knowledge of this French abbacus movement seems to have vanished while the texts themselves disappeared into dusty libraries from which only these few have recently been resurrected.

The reason for this sudden disappearance of practical mathematics in France lies, I think, in the coming of the Renaissance and the marked change that this had on the character of French intellectual life. French Renaissance humanism shared with the general humanist movement north of the Alps a theological orientation that devoted as much attention to the study of the Bible and the writings of the Christian Fathers as it did to the works of classical antiquity. This 'Christian humanism' gave even less attention to mathematics than the classical humanism of the Italians.[55] Those few French humanists who were interested in mathematics, such as Jacques Lefèvre d'Etaples (*c.* 1455-1537) and Charles de Bouvelles (1479-1566), were theologians and philosophers who saw mathematics in the light of that most mathematical of ancient philosophers, Plato. As a result, French mathematics moved away from the practical interests of the abbacists and towards the theoretical concerns of the ancients. Bouvelles published books on number theory and geometry and developed an interpretation of Plato's polyhedra;[56] Lefèvre d'Etaples' mathematical works include an edition of Euclid's geometry, an epitome of Boethius' arithmetic, and editions of Jordanus' arithmetic and music—all standard theoretical works of classical and medieval mathematics.

A similar shift away from practical mathematics characterizes the works of Oronce Fine (1494-1555), who is the true founder of mathematical studies

---

[53] Laura Toti Rigatelli (ed.), *Dionigi Gori, libro e trattato della praticha d'alcibra* (Quaderni del Centro Studi della Matematica Medioevale, 9, Università di Siena, 1984).
[54] See Barbara Moss's paper on 'Chuquet's mathematical executor' in this volume.
[55] For a further discussion of the influence of French humanism on French mathematics see Margolin, *op. cit.* (note 24).
[56] See Philip Sanders' paper on 'The platonic polyhedra' presented at this conference.

in France and the first occupant of the chair of mathematics at the *Collège Royale*. The bulk of his published works deal with astronomy and astronomical instruments;[57] his strictly mathematical works include a commentary on Euclid, a quadrature of the circle, and an *Arithmetica practica* (first published as part of his *Protomathesis* in 1532) that in fact contains only an algorism of integers, an algorism of common and sexagesimal fractions, and a treatment of ratios and proportions—a text that follows the model of the medieval algorisms. No commercial or practical problems are included. When Petrus Ramus began his campaign for educational reform in the 1540s, he complained repeatedly about the total absence of mathematical studies in the schools or the meaningless abstractions that passed for such studies, and advocated a return to the concrete examples provided by the mathematics of the marketplace and the workshop.[58] Thus it was not until the middle of the sixteenth century, under the impact of Ramus' critique, that practical mathematics and algebra again appear in France.

In view of the history just recounted, it should come as no surprise that the first algebra published in France was by a German, the Latin *Algebrae compendiosa facilisque descriptio* of Johann Scheubel, a professor of mathematics at the University of Tubingen, which was printed in Paris in 1551. This was quickly followed by a series of algebras written by French authors: the *L'algebre* of Jacques Peletier (1554), the *Logistica* of Jean Borrel [Johannes Buteo] (1559), and the *Algebra* of Petrus Ramus (1560). About two decades later Guillaume Gosselin's *De arte magna* appeared (1577), and we should also include the *L'arithmetique* of Simon Stevin (1585), which, although it was published in Leyden, was written in French and included a substantial treatment of algebra. These six works, together with the earlier works of Chuquet and de la Roche, constitute the foundation of the French algebraic tradition that was to reach such great heights in the seventeenth century.

But the most distinctive feature of this French algebraic tradition in the sixteenth century is its lack of cohesiveness, the absence of those very elements of continuity and shared development that would allow one to call it a tradition. As I have already noted, de la Roche took his inspiration as much from the Italians as he did from his master Chuquet. Scheubel's algebra was printed without alteration from a short work he had included in his edition of Euclid's *Elements* published at Basel in 1550. Its chief sources of inspiration were obviously German, and it employs the standard German cossic notation of the period. Its most significant innovation was dividing equations

---

[57] There is a list of Fine's works in Denise Hillard and Emmanuel Poulle, 'Oronce Fine et l'horloge planétaire de la Bibliothèque Sainte-Geneviève', *Bibliothèque d'humanisme et renaissance*, *33* (1971), pp. 335–351. See also Richard P. Ross, 'Oronce Fine's printed works: additions to Hillard and Poulle's bibliography', *ibid.*, *36* (1974), pp. 83–85.

[58] See Hooykaas, *op. cit.* (ref to note 24) especially Chapter X.

How algebra came to France 141

into three general classes according to whether they contained two terms, three terms with their powers in continuous proportion, or three terms in discontinuous proportion.

Peletier, who published the first true French algebra, in the sense that it is the first book devoted entirely to algebra by a French author in the French language, shows no awareness of his native forerunners, citing neither Chuquet nor de la Roche in his book. Instead he names Luca Pacioli and Girolamo Cardano from Italy, Michael Stifel, Adam Ries, and Scheubel from Germany, and Pedro Nuñez of Portugal, even though he claims to have seen neither Ries' nor Nuñez' books. But his chief source of inspiration is clearly Stifel's *Arithmetica integra* of 1543. He follows Stifel in his choice of symbolism and terminology, in using capital letters A, B, C, etc. to indicate multiple unknowns, in the use of the modern radical sign $\sqrt{}$, in the use of the word 'exponent', and in formulating a general rule for the solution of all equations instead of relying on a list of special cases. In all these choices he shows his superior sense of the qualities that were to make algebra an effective science and brought to French readers an algebra that matched the best of its time.

Yet the next work in the temporal sequence, Johannes Buteo's *Logistica* of 1559, reverted to a completely archaic set of sources and objectives. Buteo was clearly obsessed by the near reverence for all things Greek that affected so many humanistically trained scholars of the sixteenth century, as can be seen from the title of his book alone ('logistica' is the Greek word for 'practical arithmetic'). He consequently sought to build an algebra that was based on Greek (meaning geometrical) models. One result was that he did not admit the possibility of having more than three powers, since it is clearly impossible to go beyond three geometrical dimensions (or so it was universally thought at the time), and he further uses the geometrical terms *linea*, *quadraturam*, and *cubus* to designate them. Even his symbolism shows his attachment to geometry, as he draws tiny sketches of a square and a cube to represent these powers. His one symbolic innovation of note, his use of a square bracket, [, as an equals sign, comes from dividing a rectangle into two sides, [ ], and is meant as a graphic way of indicating that the following number represents an area.[59] In fact, he sometimes uses both brackets to enclose the number, as in 4ρ [12] ($4x = 12$), which reminds us that the number 12 is a rectangle, the product of two linear numbers, 4 and x.

Apart from Euclid, to whom Buteo attributes the invention of algebra in Book 2 of the *Elements*,[60] Buteo's major sources seem to have been Pacioli and Cardano. No French mathematicians are named, except for an otherwise unknown Stephanus à Rape of Lyon, who Buteo says published a

[59] p. 139.
[60] p. 7.

book on arithmetic in French.⁶¹ Gosselin also cites this Stephanus, but nothing else is known of him, at least no surviving copy of such a book is known. The geometric framework that Buteo imposed on his algebra produced a very limited and dated work, one that had no potential for future development and stimulated no successors.

Petrus Ramus was not a mathematician by training but rather a philosopher who reacted against the traditional scholastic methods of his discipline and developed a passion for educational reform, one that would make the teaching of philosophy and science more 'natural' and more closely tied to real problems and experiences. He embarked on a series of model textbooks, the *Scholae*,⁶² which were meant to illustrate his ideas of how the liberal arts should be taught, and was led to include mathematics because of the essential role that it must play in any educational system. In addition to major texts on arithmetic and geometry, he published a small book of thirty pages on algebra which appeared anonymously in Paris in 1560. The fact that it was published anonymously is generally taken as an indication that Ramus was somewhat uncertain of the role that algebra should play in mathematics. As a 'new' field of mathematics, one that had not been mentioned by Plato or Aristotle and had no classical models (he may have been unaware of Diophantus' work at this time), Ramus was uncertain about where it fitted into his new scheme of education. In the book itself he makes algebra a part of arithmetic and ties it to the theory of proportions, calling it 'that part of arithmetic that sets up its own proper numbers from continuously proportional figures',⁶³ where 'figures' are algebraic numbers, i.e. unknowns and their powers. The motivation for this can be seen from the fact that the scale of powers can be constructed from a continued proportion, that is $1:x:x^2:x^3:x^4:x^5:\ldots$ However, Ramus retains a common geometrical terminology for the powers, calling them *latus*, *quadratus*, *cubus*, *biquadratus*, *solidus*, etc., and his symbols are simply the first letters of these words. Like Buteo, Ramus divides equations into two major types: simple equations with two terms equal to one another, and complex equations with three terms, two equal to one. The latter are divided into three canons, but the use of this word should not be taken to indicate any awareness of Chuquet's work, since the same classification and terms were used by Buteo and by Scheubel before him, and the division into simple and complex equations goes back to al-Khwarizmi. Ramus gives no indication of the sources he used.

Guillaume Gosselin is the only one of the French algebraists to cite the

---

⁶¹ p. 7. Also cited on pp. 118, 183, and 184.
⁶² There is a list of Ramus' many works in Walter J. Ong, *Ramus and Talon inventory* (Cambridge: Harvard University Press, 1958).
⁶³ '*Algebra est pars arithmeticae, quae e figuratis continue proportionalibus numerationem quandam propriam instituit.*' (f. 2ʳ).

What could we learn from Master Christianus van Varenbraken? A note on an arithmetic manuscript of a sixteenth-century Flemish[1] schoolmaster

MARJOLEIN KOOL

THE author of an arithmetic manuscript, written in about 1532, was probably Christianus van Varenbraken. This paper describes this manuscript and compares it with other sixteenth-century Dutch arithmetics.[2]

1. Early sixteenth-century Dutch arithmetics

The first printed arithmetic in the Dutch vernacular is that printed by Thomas Vander Noot at Brussels, in 1508, *Die Maniere om te leeren cyffren na die rechte consten Algorismi int gheheel ende int ghebroken* (the way of learning to calculate according to the true art of Algorisms in whole and broken numbers). An edition of this arithmetic, enlarged with a part about reckoning with counters, was printed in about 1510 by Willem Vorsterman at Antwerp. It is not impossible that part of the arithmetic of 1508 was reprinted by Jan Seversz in 1527. The authors of these books are unknown. The next printed Dutch arithmetic, written in the vernacular, was *Een sonderlinghe boeck in dye edel conste Arithmetica* (an extraordinary book on the noble art of arithmetic) written in 1537 by Gielis Vanden Hoecke. This arithmetic was reprinted in 1545. Finally, there are some indications of another Dutch arithmetic, written in the vernacular and printed in the first part of the sixteenth century, but the existence of this book is not certain.[3]

As far as is known, these above-mentioned books are the only arithmetics printed in the Netherlands during the first half of the sixteenth century that

---

[1] The germanic language of the medieval Low Countries is called 'Middle Dutch'. Flemish is one of the principal dialects of this language. It is the dialect of Flanders.
[2] I am working on an edition of the arithmetic, to appear (1988) in the *Scripta* series of studies and editions of mediaeval and Renaissance texts published by the Research Center of Mediaeval and Renaissance Studies (OMIREL) of the UFSAL in Brussels.
[3] Smeur, 1960, p. 13. (See Bibliography at the end of this paper.)

were written in Dutch. During the same period, many arithmetics in French and Latin were printed in the Netherlands, most of them at Antwerp. Especially notable is an *Arithmeticae practicae methodus facilis*, written by Gemma Frisius in 1540. In contrast to the earlier part of the sixteenth century, many Dutch arithmetics written in the vernacular were printed in the second half of that century. They were written by Creszfelt (1557), Raets (1566), Petri (1567), Vander Gucht (1569), Helmduyn (1569) and others.

At present little is known about arithmetics in the Dutch vernacular in the early sixteenth century, their intended public, and the pedagogical principles used in their composition. I hope that this paper contributes to the research on these subjects.

2. A sixteenth-century Dutch arithmetic manuscript

In the University Library of Ghent in Belgium is a Dutch arithmetic manuscript to which, until recently, little attention was paid. The arithmetic contains about 130 pages and is part of a paper codex (folios 127^r–191^r). This codex (signature 2141) was bought by the University Library of Ghent in 1913. The history of the codex before 1913 is unknown.[4]

Some other parts of the codex are

- a tract on *orthographia* or *scriftwere* (folios 4–32).
- a didactical poem on the Dutch spelling (folios 34–43).
- early seventeenth century geometrical tracts (folios 65–76). They were probably written by a certain De Buele and added later than the other parts of the codex.
- a tract on music (folios 91–121). This tract is the oldest work on music in the Dutch vernacular.
- a tract on geometry (folios 192–260). This is probably a copy of the first printed Dutch work on geometry, *Die waerachtige conste der geometrien* (The true art of geometry), 1513.

The arithmetic and most of the other tracts in the codex were written by the same person, whose name can be found at several places in the text. It is Christianus van Varenbraken. Perhaps he is the author or at any rate the scribe of the text. Christianus writes that he is a master in the liberal arts and that he has taught reading and writing to many people. Nothing else is known about this man.

On folio 185^r of the arithmetic can be read:

---

[4] For further information on the parts of this codex, see Bockstaele, 1984 and Braekman, 1978 and 1981.

Finitur per me Christianus van Vanderbraken practicus in omnibus artibus liberalis ipso die post festium Sancti Johannis anno domini 1532.

Probably the arithmetic manuscript was written in 1532. The text may have been written for merchants, because in the prologue of the arithmetic can be read: 'This text deals with *arithmetica naer der coopmanscepe*', that is, 'arithmetic for matters of business'.

3. The contents of the arithmetic manuscript

3.1 The operations (127ʳ–146ʲ)

The arithmetic consists of three distinct parts. The first 40 pages deal with the *species algorismi*. These are: numeration, addition, subtraction, halving, doubling, multiplication and division. In *Die Maniere...* (1508 and 1510?), the number of operations dealt with is also seven, just like in some of the Latin sixteenth-century arithmetics. However in Vanden Hoecke (1537) and most of the other sixteenth-century arithmetics the number of operations dealt with is four (addition, subtraction, multiplication and division. Numeration is also treated, but no longer considered an operation.) The number of seven operations is somewhat old-fashioned in a sixteenth-century arithmetic.

For each separate operation, the text explains how one must calculate with integers, directly followed by an explanation of that same operation with fractions. As far as is known, this is the only arithmetic which follows this order. The usual order is: first an explanation of *all* the operations with integers, followed by an explanation of *all* the operations with fractions.

The arithmetic starts with a short prologue: an eulogy on arithmetic, the foundation of the other arts. After that, in the section on numeration, one can learn to read and write Hindu–Arabic numerals. The numbers are classified in three classes: *digiten* (unities), *articulen* (tens) and *compositen* (composite numbers). The author already knew the term 'million' but he uses this term only once. He writes that he prefers to use the name 'thousandthousand'. So he uses 'one-thousandthousand', 'two-thousandthousand', and so on. In the part about numeration with fraction, the terms 'numerator' (in Dutch, *teller*) and 'denominator' (in Dutch, *noemer*) are used. Fractions are denoted by writing the numerator above the denominator, separated by a mark of division. In *Die Maniere...* of 1508 the mark of division is not obligatory; the reader may omit it, if he chooses.

The explanation of addition starts with a difficult example in denominated numbers. Denominated numbers are numbers composed of several currencies. It is not clear why the author starts this part with a difficult example, and after that a much easier one. Folio 130ʲ:

150   *Mathematics in the sixteenth century*

Ian   ⎫   ⎧32h 12s 9d 16te   h = ponden
Pieter   ⎬ owes to me   ⎨47h 17s 10d 18te   s = scelling
Willem   ⎪   ⎪28h 19s 11d 19te   d = penningh
Wouter   ⎭   ⎩24h 16s 8d 22te   te = miten

Question: How much is owed to me?
  1h = 20s = 240d = 5760te.

The result is checked by casting out nines, like most of the examples in this arithmetic. The method of casting out nines works as follows: first one must divide the result of the addition by 9, after that one must divide each term of the addition by 9 and add up the remainders, then one must divide the result of that addition by 9. If no mistakes are made, the last remainder will be the same as the remainder of the first division.

The part about addition with fractions starts with an example containing fractions that have the same denominator, and after that an example containing fractions with different denominators. To make these fractions of the same denomination, it is necessary to use multiplication. This operation, however, has not yet been explained. This example demonstrates that the author, by dealing with the integers first and with the fractions second, at some places needs operations he has not taught yet. There are no places in the text showing that he has any objections against this, and that is rather curious.

In contrast to addition with integers, the order of the section on subtraction is, pedagogically speaking, more straightforward. This part starts with a simple example, followed by a more difficult one: subtraction with borrowing. In this arithmetic, the same method of subtraction is used as we use nowadays.

In this context a peculiar sentence attracts attention. Folio 134^r:

But if these numbers are odd or if the lower number in principle is more than the upper number, you have to borrow . . .

The second part of this sentence is not wrong, but the first part is very strange in this context. The difference between odd and even numbers has nothing to do with subtraction. In the operation halving, it plays a role and indeed in the text about halving the same words are repeated (folio 137^r). In the text about subtraction this sentence must be a mistake. *Die maniere* . . . of 1510(?) contains exactly the same mistake. *Die maniere* . . . of 1508 is correct at this place: the sentence is omitted from the text on subtraction and appears only in the part on halving. This enables us to say more about the relations between the manuscript of Christianus van Varenbraken and the two editions of *Die maniere*. . . . In contrast to *Die maniere* . . . of 1508, *Die maniere* . . . of 1510(?) contains the same mistake as the manuscript of Christianus van Varenbraken. It follows that the arithmetic of Christianus van Varenbraken is more closely related to *Die maniere* ... of 1510(?) than to *Die Maniere* ... of 1508.

The arithmetic continues with halving and doubling. To us these operations seem superfluous because halving is the same as division by 2 and doubling is the same as multiplication by 2.

The next part, about multiplication, starts with a 12 by 12 table of multiplication. After that, some examples of multiplication follow. The way of multiplication is nearly the same as in the modern method, with but one exception: during the process, figures that have been used and that are no longer necessary for calculation are ticked off. When all the figures have been ticked off, the multiplication is finished. But then it is impossible to read back the calculation in order to recalculate it. So other ways of checking are needed. In this arithmetic the calculation is checked in two ways: casting out nines and the inverse operation: division.

The method of division explained in this arithmetic is the *divisio galea* or *divisio batello*, so called because after dividing, the sum looked like a sailing-ship.[5] In this arithmetic, the three numbers of a division are placed below one another in the following order:

> dividend
> quotient
> divisor

As in multiplication, in division the used figures are ticked off, and here also two ways of checking are given: the 9-proof and the inverse operation: multiplication. Division is the last operation dealt with in the first part of the arithmetic.

3.1.1. Some peculiar aspects of the section on the operations

There are some striking aspects of the presentation of the seven operations in this arithmetic. In the first place it is remarkable that this arithmetic only gives rules and examples to practise these rules. It is just like a cookery-book; the arithmetic only gives recipes and never an explanation why it is necessary to do it in that specific way. In this respect there is a great difference between this arithmetic and our modern textbooks.

Another curious aspect, which has been mentioned above, is the way in which the subjects are ordered in this arithmetic: in each separate operation, first the integers are treated, directly followed by the fractions. As mentioned before, this order causes problems, for example in the addition of fractions that have different denominators.

Another strange aspect is the pedagogical composition of some parts of this arithmetic. For example, in the part about addition, the explanation of

---

[5] cf. Jackson, 1906, pp. 69–70 and Tropfke, 1980, pp. 236–239.

152 *Mathematics in the sixteenth century*

addition starts with a difficult example in denominated numbers. After that a much easier example follows.

A final peculiar aspect: there are some aspects of arithmetic to which hardly any attention is paid in this arithmetic. For example, 'reducing fractions' is explained in only one sentence, with one example, and that is all.

3.2 *Rules and problems (folios 146^v–185^r)*

The second part of the arithmetic consists of about eighty pages, and contains rules and a collection of problems. This part is an application of the things learnt in the first part. Each time a rule has been treated, some problems solved by this rule follow, and sometimes there are also other problems which are solved without this rule. Some of the problems are practical and deal with buying and selling of all kinds of goods with different prices, weights and measurements. Other problems are more like puzzles to solve for pleasure. Sometimes it is difficult to draw the line between these two kinds of problems.

The first rule dealt with in this part is the rule of three. This rule takes three given numbers and provides a method for finding a fourth number which stands in the same proportion to the third as does the second number to the first. An example of a problem solved with the rule of three is: Given: 100 pounds of wax cost 4 pounds. How much is 25 pounds of wax? To solve this, one must put the three numbers next to one another on a line: 100----4----25. To find the required number, one has to multiply the last two numbers and after that to divide the result by the first number. Here: 4 times 25 makes 100, and 100 divided by 100 makes 1. So 25 pounds of wax cost 1 pound. The rule of three is the most important rule of the sixteenth-century arithmetic; this rule is also called the 'golden rule'. The other rules dealt with in this arithmetic are in principle the same as the rule of three, but they have different names which derive from the contexts in which they are used. For example: to solve a problem about changing money, the rule of three is necessary. In the contents of exchange problems, this rule is called the 'rule of exchange'. In the same way, there is a rule of company, a rule of bartering, and some rules with names that refer to one of the seven operations, for example, rule of halving, rule of doubling, rule of multiplication, rule of division, and so on. First, all the rules on integers are dealt with, and after that the rules on fractions. As in the first part, this part of the arithmetic does not aim at giving real insight into arithmetic, but at giving the reader a mechanical recipe.

Most of the problems of this arithmetic are also found in other sixteenth-century arithmetics. Among the problems in this arithmetic, there are some problems that have nothing to do with these rules. For instance, the old problem of the two men who must divide 8 pints of beer, using only an 8-pint can, a 5-pint can and a 3-pint can. How can they get two equal parts of beer (folio 149^v). Or the problem at folio 177^v:

What could we learn from Master Christianus van Varenbraken? 153

It takes a drunkard 14 days to empty a barrel of beer by himself, and when his wife is drinking together with him, they empty a barrel in 10 days. The question is: how much time does his wife take to empty the barrel when she is drinking alone?

These two problems also illustrate the recreational problems that were mentioned above.

Three pages of this part of the arithmetic are in Latin. On these pages, the *regula de falsis* is dealt with. In the two editions of *Die Maniere . . .* this rule is not treated. These two arithmetics contain not a word in Latin. Possibly the author of the manuscript has used a Latin source for this part.

3.3 Reckoning with counters (folios 185ʳ–191ʳ)

Part three of the arithmetic is about 10 pages long and contains an explanation of reckoning with counters. This is a way of calculating developed from line reckoning.[6] In this method, the lines are replaced by counters that indicate unities, tens, hundreds and so on. This part was added to the arithmetic for all the people that could not read and write. It explains the seven operations in reckoning with counters in some examples. The arithmetic ends with four simple problems.

In *Die maniere . . .* of 1508 nothing is explained about reckoning with counters. In *Die maniere . . .* of 1510(?), however, the part on reckoning with counters is exactly the same as in the arithmetic manuscript, which is another indication of the close relationship between these two arithmetics. The arithmetic of Vanden Hoecke contains a section on line reckoning.

4. The significance of the arithmetic manuscript

From a comparison of the early sixteenth-century vernacular arithmetics one can conclude that there are many similarities between them. Both the 1532 manuscript and the arithmetic of Vanden Hoecke cover much of the same material as in the first two editions of *Die maniere . . .*, although these two later arithmetics are themselves quite different. The arithmetic of Vanden Hoecke is far more comprehensive than the earlier arithmetics; it contains sections on the extraction of roots, progression, proportion and algebra. The arithmetic manuscript of 1532 is closely related to the first two editions of *Die maniere . . .*; its resemblance to the 1510(?) edition of *Die maniere . . .* is particularly striking. Both arithmetics have literally the same passage on reckoning with counters, and sometimes the very same mistakes are found in the two texts, as in the above-mentioned sentence in the section about subtraction. However, the 1532 manuscript contains more problems than are provided in *Die maniere . . .* of 1510(?). It is not impossible that the author has used other sources for these problems, because most of them also occur in other sixteenth-century arithmetics. Why should an author add 'new'

---

[6] cf. Jackson, 1906, pp. 25–26 and Smeur, 1960, pp. 124–127.

problems to an arithmetic like *Die Maniere* . . . of 1510(?)? Is it to improve his teaching?

This question is an aspect of the main question of this study: what can we learn about the pedagogical principles of Christianus van Varenbraken and the public for which he wrote? To answer this question, it is important to compare this arithmetic with other arithmetics. In this comparision, it is especially important to pay attention to the above-mentioned peculiar aspects of this manuscript:

- The author has changed the usual order of the subjects in his arithmetic.
- There are subjects of arithmetic to which hardly any attention is paid.
- The arithmetic contains many practical problems from what was known as 'arithmetic for matters of business'.
- The problems are presented as a matter of rote learning.

Research on these aspects in particular in other arithmetics may possibly bring us into further contact with master Christianus van Varenbraken and his pupils.

BIBLIOGRAPHY

Primary sources

1508 *Die maniere om te leeren cyffren na die rechte consten algorismi, int gheheel ende int ghebroken.* Brussels, Thomas Vander Noot. Royal Library, Brussels and British Museum, London. (Nijhof and Kronenberg, 1923–71, no. 1482.)

1510? *Die maniere om te leeren cyffren ende rekenen metter pennen en metter penningen na die gherechte conste algorismi. Int gheheele ende int ghebroken.* University of Amsterdam Library (ned. inc. 293). (Burger, 1929.)

1513 *Die waerachtige const der geometrien* . . . Brussels, Thomas Vander Noot. University Library, Ghent (R. 1099) (Bockstaele, 1984.)

1527 *Kalengier ende die maniere om te leeren cyffren.* Antwerpen, Jan Seversz. te koop bij dezen, aldaar, 1527. This is probably a reprint of a part of the arithmetic of 1508. It is not known where the only copy of this book of 1527 is at this moment. Until 1940 it was in the possession of Mr Jervis Wegg in London. Since then it has been sold several times. (Nijhof and Kronenberg, 1923–1971, no. 3295, see also Smeur 1960, p. 11).

1532 An arithmetic manuscript. Folios 127^r–191^r of codex 2121, University Library, Ghent.

1537 G. Vanden Hoecke. *Een sonderlinghe boeck in dye edel conste arithmetica.* Antwerp, Symon Cock. Plimpton Library, New York. (Nijhof

and Kronenberg, 1923-71, no. 3175.)
1545 Reprint of the above. Antwerp, Symon Cock. Royal Library, Brussels, and British Museum, London.

Secondary sources

Bockstaele, P. (1959) 'Het oudste gedrukte Nederlandse rekenboekje', *Scientiarum historia*, *1*, pp. 53-71.

Bockstaele, P. (1960) 'Notes on the first arithmetics printed in Dutch', *Isis*, *51*, pp. 315-321.

Bockstaele, P. (1984) 'Het oudst bekende, gedrukte Nederlandse meetkundeboek: Die Waerachtige const der geometrien', *Tijdschrift voor de geschiedenis der geneeskunde, natuurwetenschappen, wiskunde en techniek*, 7, pp. 79-92.

Braekman, W.L. (1978) 'Twee nieuwe traktaten uit de vroege zestiende eeuw over de Nederlandse spelling', *Verslagen en mededelingen van de kon. Academie van Nederlandse Taal- en Letterkunde*, 1978, pp. 294-306.

Braekman, W.L. (1981) 'Christianus van Varenbrakens Conste van musycke oft vanden sanghe', *Scripta*, 5.

Burger, C.P. (1929) 'ABC-penningen of rekenpenningen', *Het Boek*, *18*, pp. 196-202.

Jackson, L.L. (1906) *The educational significance of sixteenth-century arithmetic*, New York.

Nijhof, W. and M.E. Kronenberg, (1923-71) *Nederlandsche bibliographie van 1500-1540*. 's Gravenhage.

Smeur, A.J.E.M. (1960) *De Zestiende-eeuwse Nederlandse rekenboeken. (The sixteenth-century arithmetics printed in the Netherlands, with summary in English.)* 's Gravenhage.

Tropfke, J. (1980) *Geschichte der Elementarmathematik*. 4th edition, edited by K. Vogel et al. Berlin and New York.

Struik, D.J. (1936) 'Mathematics in the Netherlands during the first half of the sixteenth century', *Isis*, *25*, pp. 46-56.

A note on Rudolf Snellius and the early history of mathematics in Leiden

K. VAN BERKEL

ACCORDING to the accepted view, which I do not intend to criticize in this paper, the development of early modern mathematics occurred mainly outside the university system.[1] Nevertheless, there are some notable exceptions to this, one of them being the flourishing mathematical school in Leiden around the middle of the seventeenth century. At that time, Frans van Schooten the younger, assisted by some of his students (men like Christiaan Huygens, Johan de Witt, Johannes Hudde and Hendrick van Heuraet) did excellent work in the new mathematics of Descartes. It is well known that contemporary mathematicians lacking proficiency in French learned Descartes' mathematics from van Schooten's Latin edition of the *Géométrie* in 1649.[2]

However, the work of van Schooten was only the culmination of a mathematical tradition in Leiden that had developed from the end of the sixteenth century. To understand why mathematics became so important in Leiden, we have to go back to the late sixteenth century, to the work of the first professor of mathematics in Leiden, Rudolf Snel van Royen, better known as Rudolf Snellius. It is interesting to see the circumstances under which mathematics was incorporated into the university curriculum.

Rudolf Snellius was born in the little town of Oudewater in the southern part of Holland in 1546.[3] He went to school in Utrecht and studied mathematics and Hebrew in Germany. He visited Wittenberg, Heidelberg and Marburg, where he became a *magister artium* in 1572. After that, he learned

---

[1] As far as the English universities are concerned, this thesis is criticized by M. Feingold, *The mathematicians' apprenticeship. Science, universities and society in England, 1560–1640* (Cambridge University Press, 1984).

[2] On Van Schooten, see the *Dictionary of scientific biography, in voce*. Mr van Maanen from the University of Utrecht (Holland) is preparing a thesis on van Schooten and his pupils. See: J. A. van Maanen, 'Hendrick van Heuraet (1634–1660?): his life and mathematical work', in *Centaurus*, 27 (1984) pp. 218–279.

[3] For all the information concerning Snellius, see: K. van Berkel, *Isaac Beeckman (1588–1637) en de mechanisering van het wereldbeeld* (Amsterdam, 1983) (in Dutch, with a summary in English), especially pp. 271–279.

sufficient medicine to get a degree in some Italian university. He returned to Germany, where he lectured *totam cyclopediam* in Marburg, one of the few Calvinist universities in that country. During his stay in Germany, Snellius had become fascinated by the philosophy of Petrus Ramus and in Marburg he lectured on it. His students were so pleased with these lectures that they published them without his knowledge or consent. Only later did Snellius revise these 'pirate editions' for republication.[4]

In 1575 Snellius returned to Holland, where he became a physician in his native town of Oudewater. He probably would have stayed there for all of his life, had not some students from Leiden asked him in 1579 to give some lectures on mathematics in the university. There had been no opportunity to study mathematics at the university since its foundation in 1575. However, the curators of the university were very reluctant to give Snellius any sort of official position in Leiden. In 1580, they officially gave him permission to teach mathematics, but on the condition that he would be dismissed as soon as another mathematician turned up who had more experience or a better reputation. This was not very encouraging, of course, but Snellius managed to stay. On 2nd August 1581, the curators appointed him 'extraordinaris professor' in mathematics.

Why were those in charge of the university so reluctant to admit Snellius and his mathematics? One reason may have been the contempt of the humanists for mathematics, which was associated with the crafts. But the main reason must have been their dislike of Snellius' Ramism. Snellius admired Ramus very much, both as a school reformer and as a mathematician. Snellius agreed with Ramus' view that the teaching of almost everything in the universities was much too complicated, too theoretical, too abstract. In his opinion, both logic and the other liberal arts, including mathematics, had to be made more simple, more concrete, more practically orientated. The reason for the study of all philsophy resides in its 'use'. '*Origo et initium philosophiae*', Snellius wrote, '*est ab usu; finis philosophiae est in usu; philosophia ipsa tendit ad vitae humanae usum ac fructum.*'[5] Ramus had boasted of his familiarity with all the shops and market-places in Paris, where mathematics was put to use. He held that only those parts of current mathematics were worth teaching that had a demonstrable practical application. Mathematics as studied in the universities should not be more than a systematic treatment of the mathematical methods in use by merchants, navigators, surveyors and engineers.

Snellius, however, was not as dogmatic as Ramus had been; his attitude to

---

[4] K. van Berkel, 'De geschriften van Rudolf Snellius. Een bijdrage tot de geschiedenis van het ramisme in Nederland', in: *Tijdschrift voor de geschiedenis der geneeskunde, natuurwetenschappen, wiskunde en techniek*, 6 (1983) pp. 185-194 (with a summary in English).
[5] R. Snellius, *Snellio-Ramaeum philosophiae syntagma* (Frankfurt, 1596) p. 76.

the practical arts was more positive. Ramus went to the workshops to see what part of mathematics was *not* used in practical daily life. He did not think it possible to learn something really new thereby. By contrast, Snellius thought it possible to obtain new knowledge by looking at the work of merchants, blacksmiths and musicians: knowledge that was not already contained in the classical texts.

Consequently, it is easy to see why Leiden University was reluctant to admit Snellius and why they kept him at a distance by making him 'extraordinaris professor' only. Ramism as taught by Snellius was a plea for breaking down the social and intellectual barriers between the theoretical science inside the university and the practical arts outside.

Even after Snellius was admitted as an 'extraordinaris professor', some of his colleagues did not treat him very kindly. Outside the university, for instance in committees for studying all kinds of patents, Snellius was treated with respect, but leading professors such as Lipsius and Scaliger had little respect for Ramus and Snellius. Lipsius once wrote to a student: 'Young man, listen to me: you will never be a great man if you think that Ramus was a great man'. And Scaliger once had a quarrel with Snellius when Snellius had had the courage to point out that Scaliger had made some elementary mathematical mistakes in his book on the quadrature of the circle.

In his lectures, Snellius was accustomed to use some of the works of Ramus, but when in 1591 he asked permission to reduce the number of his courses, he was only allowed to do so on condition that he stopped reading Ramus and read instead a classical author such as Euclid or Aratus. But this did not stop him from propagating Ramist mathematics. We know of a list of suggested reading he prepared for one of his students in 1607, on which the works of Ramus and his followers had a prominent position.[6]

It is also clear from the list of his publications that Snellius continued to spread the ideas of Ramus. He wrote compendia for all kinds of sciences, from ethics to arithmetics, from dialectic to geometry and from physics to psychology, all of which exhibited the Ramist approach. The fact that most of his books were published in Germany suggests that his reputation there may have been higher than it was in Leiden.[7] In Leiden, Snellius was quite popular with the students, but at best only tolerated by the humanists and the curators.

In 1600, a change took place whose ramifications were crucial for the history of Dutch mathematics. In January 1600, Maurits, Prince of Orange and Stadholder of Holland and Zeeland, founded a school for military

---

[6] Snellius gave this list to Isaac Beeckman. See: *Journal tenu par Isaac Beeckman de 1604 à 1634, publié avec une introduction et des notes par C. de Waard* (4 vols., The Hague, 1939–1953), especially vol. IV, pp. 17–19. For the dating of this list and a commentary on it, see: Van Berkel, *Isaac Beeckman* . . ., pp. 25–26, 281–282.

[7] See note 4.

A note on Rudolf Snellius and the early history of mathematics in Leiden   159

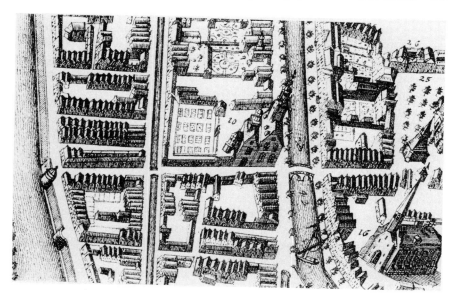

FIGURE 1. The University of Leiden in the second half of the seventeenth century. Detail of a map of Leiden by Christiaan Hagen, dated 1675.

This detail from Hagen's map shows some university buildings in the southwestern part of Leiden (the north is at the top of this picture). The building indicated by the number 10 is the *Academiegebouw*, the official seat of the university since 1581 and the place where the lectures were given. On top of the roof of the *Academiegebouw* there are two towers, the smaller of the two being the astronomical observatory founded in 1633 by Jacob Golius. Behind the *Academiegebouw* we can see the Botanical Garden, laid out in 1594.

Close to the *Academiegebouw*, on the other side of the canal (*Het Rappenburg*), we see (number 16) the *Faliede Bagijnenkerk* or *Engelse Kerk* (English Church). From 1577 to 1581 this building was the seat of the university; after 1591 it was used for the university library (on the first floor) and the Anatomical Theatre (in the absis). After 1594, Ludolf van Ceulen was allowed to give his lectures in the art of fencing on the ground floor of the church, and in 1600 the building was partly used for the Mathematical School. The different uses to which the same building was put simultaneously illustrates the close links between the university and the Mathematical School.

The *Academiegebouw* is still standing, but the observatory was moved in the nineteenth century. The canal is nowadays one of the most beautiful parts of Leiden. The *Faliede Bagijnenkerk* is still there too, but it has been altered and extended to such an extent that it has become unrecognizable today. A few years ago, the university library moved to another location in Leiden.

engineers in Leiden.⁸ Simon Stevin drew up the statutes and the curriculum for this new institution. An important innovation was that the lessons, intended for 'illiterate' masons, navigators, carpenters, surveyors and engineers, were given in the vernacular, not in Latin. In winter, the students were in Leiden, but during the summer, those who intended to become military engineers joined the army to learn their craft in the field. The main part of the lectures was devoted to mathematics, both theoretical and practical, and that is why this school was called *De duytsche mathematycque*, the Dutch Mathematical School. One of the first instructors was Ludolf van Ceulen, famous for his determination of the number π.

This new mathematical school was not part of the university, but it had close connections with it. The school was housed in a university building and its students and teachers had access to the university library (housed in the same building). Ludolf van Ceulen had been appointed as a fencing-master for university students in 1594 and he continued in this capacity after his appointment to the Mathematical School in January 1600. Both he and his colleague Simon van Merwen held the title of lecturer.⁹

The creation of this new institution had a considerable influence on the teaching of mathematics in the Dutch universities. The first to react was the University of Franeker, founded in 1585. The beginnings of mathematics in Franeker were as difficult as in Leiden.¹⁰ The first professor was someone called Roggius. He was a lecturer in Divinity who could not keep his students in order and was therefore in 1594 'promoted' to the chair of mathematics. However, Roggius was not at all equipped for this and he left in 1596. Only two years later, in 1598, a new mathematician was appointed, Adriaan Metius, son of Adriaan Anthonisz, a burgomaster of Alkmaar and a military engineer in the service of the stadtholder, Maurits. Metius had studied with Snellius in Leiden and Tycho Brahe at Hven. As with Snellius, Metius was only appointed as an 'extraordinaris professor', but on 22 April 1600, a few months after the foundation of the mathematical school in Leiden, he was made a full professor. Metius lectured both in Dutch and Latin on abstract

---

⁸ There exists no detailed study of the Mathematical School in Leiden. Information on the school can be gathered from: P.C. Molhuysen, ed., *Bronnen tot de geschiedenis der Leidsche universiteit*, I (Leiden, 1913), especially pp. 389*-392*, 411*-412*; E. Taverne, *In 't land van belofte: in de nieuwe stadt. Ideaal en werkelijkheid van de stadsuitleg in de Republiek 1580-1680* (Maarssen, 1978) pp. 61-66.

⁹ Van Merwen died in April 1610, Van Ceulen in December 1610. Their courses were taken over by Frans van Schooten the elder, who was appointed *lector Matheseos Teutonicae Linguae* in 1612. He became professor of mathematics in 1615. After his death in December 1645, he was succeeded by his son Frans van Schooten the younger.

¹⁰ On Roggius and Metius, see: W.B.S. Boeles, *Frieslands hoogeschool en het Rijksathenaeum te Franeker* (2 vols, Leeuwarden, 1878-1879), especially vol. II, pp. 62-66, 70-75. On Metius: H.A.M. Snelders, 'Alkmaarse naturwetenschappers uit de 16de en 17de eeuw', in: *Van Spaans beleg tot Bataafse tijd. Alkmaars stedelijk leven in de 17e en 18e eeuw* (Zutphen, 1980) pp. 101-122, especially pp. 105-108.

mathematical theory and practical mathematics to navigators, surveyors, engineers and university students alike. Therefore, he combined two tasks that were separated in Leiden, but his promotion in 1600 was certainly influenced by the developments in Leiden.

Once the Mathematical School in Leiden had appointed two lecturers in mathematics and Franeker University had promoted its 'extraordinaris professor' to a full professorship, the curators of Leiden University could not remain backward. Therefore Snellius, who had close connections with the Mathematical School (he was a member of the board of examiners), became a full professor in mathematics in February 1601, a year after the courses in the new Mathematical School had begun. Although we have no conclusive evidence for this connection, it cannot be a coincidence that Snellius' promotion took place within a year after the beginnings of the Mathematical School.

What kind of conclusion can we draw from this? It is a commonplace in the history of universities that newly founded universities are more open to new intellectual movements than older ones. There is some truth in it, and numerous examples can be adduced. Leiden University was much more open to the new humanistic learning than many older universities in France or in Germany. But we have also seen that there exist counter-examples. The early history of mathematics in Leiden does not show us how mathematics was *welcomed* in Leiden, but how it was *forced* upon a reluctant university. In late sixteenth-century Holland, the practical mathematics Snellius propagated inside the university dominated scientific life outside the university, but the curators only created a restricted space for it in the curriculum. Only after the institutionalization of practical mathematics in the new school designed by Stevin and the promotion of Metius in Franeker did Leiden University comply.

This concession to the practical mathematicians was the beginning of a flourishing mathematical tradition. Rudolf Snellius may not have been a great mathematician, but his successors Willebrord Snellius and Jacob Golius certainly were. Similarly, the fame of Frans van Schooten the younger is undisputed. Although he taught at the Mathematical School, he is mainly remembered for his work on Cartesian mathematics and for his stimulating influence on some excellent university students. By the time Van Schooten published his Latin translation of Descartes' *Géométrie*, mathematics had cast off its practical associations, but, as I have tried to show, they had been very important in the institutionalization of mathematics half a century before.

The first arithmetic book of Francisco Maurolico, written in 1557 and printed in 1575: a step towards a theory of numbers

JEAN CASSINET

FRANCISCO Maurolico was born and died at Messina in Sicily (19 September 1494–21 July 1575). His parents were Greeks from Constantinople who had fled this town to take refuge in Messina in 1453, shortly before the capture of Byzantium by Mohammed II. It was Maurolico's father who taught him mathematics.

Maurolico's work in arithmetic[1] is contained in its entirety in an *Arithmetic* in two books, printed in 1575, that is, after the death of the author, but written in 1557. The first volume was finished on Easter Sunday 18 April 1557 at three o'clock in the morning, the second on 24 July in the same year at 6 p.m., as the colophons placed at the end of each of these books of the printed work testify.

Maurolico's place as an arithmetician is important in an epoch when the first Latin translation of the work of the great arithmetician of antiquity, Diophantus of Alexandria, also appeared; indeed, it is known that it was also in 1575 that the edition due to Xylander appeared, an edition which was shortly to be followed by the well-known ones of Bachet de Méziriac in 1621, and of Fermat in 1670.[2] The conjunction in 1575 of the appearance of the arithmetic of the Abbot of Messina and of the first Latin edition of Diophantus constituted an important element in the beginnings of interest in the theory of numbers which would make its way by Bachet de Méziriac and above all by Fermat at the beginning of the seventeenth century. Moreover, Maurolico was cited by Bachet and also by Blaise Pascal.

---

[1] F. Maurolycus, *Arithmeticorum libri duo*, Venice, 1575.

[2] Diophantus of Alexandria, 'Diophanti Alexandrini Libri sex' followed by 'Librum de Numeris polygonis'.
1575 edition, due to Xylander, Latin text only.
1621 edition, due to Bachet de Méziriac, bilingual Latin–Greek, with numerous commentaries by Bachet.
1670 edition, due to S. de Fermat, based on the edition of 1621 with additional notes by P. de Fermat.

The first arithmetic book of Francisco Maurolico 163

Bachet de Méziriac cites him in one of his notes which appeared in the 1621 edition concerning proposition 27 in Book II in the Appendix *de Multangulis Numeris*.[3] The subject was a proof by Bachet of a long-established theorem on the sums of consecutive odd numbers which yield the sequence of cubes

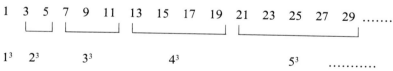

Pierre Fermat himself did not cite Maurolico explicitly, but it is known that his marginal notes suffered from an exiguity of which he frequently complained. However, concerning this same theorem, he wrote a note which appeared, to be specific, on page 40 of the *de Multangulis numeris* of Diophantus which is found after the Arithmetics of Diophantus; there Fermat generalizes proposition 27 (question 62 in Maurolico) by making use of the 'columns' nomenclature which appears in Maurolico and which the latter invented, contrary to what Paul Tannéry thought and wrote in his 1891 edition of the works of Fermat, where he attributed its invention to the native of Toulouse.[4] The use by Fermat of terminology and categories characteristic of Maurolico shows clearly that he knew the arithmetical work of the Abbot of Messina.

As for Blaise Pascal, it was in his *Histoire de la roulette*, published in 1659, that he wrote concerning a proposition stating that twice a triangular number diminished by its index is equal to the square of this index $(2 \times \frac{n(n+1)}{2} - n = n^2)$, that he wrote:[5]

Cela est aisé par Maurolic

In fact, Pascal means here that this was facilitated by induction, which supports the assertion—perhaps somewhat exaggerated—of the historian of mathematics G. Vacca who, in an article written in 1910,[6] called Maurolico the discoverer of the principle of Mathematical Induction.

All this emphasizes the role played by Maurolico in the interest in the theory of numbers which awakened in several seventeenth-century authors, even if it is not until the end of the eighteenth century that a real theory of numbers is seen to take shape.

---

[3] 'Scholium: Eadem fere ratione demonstratur haec propositio a Francisco Maurolyco, prop. 62 arithmeticorum', from the 1670 edition of 'de Multangulis Numeris', p. 39.

[4] P. Fermat, *Oeuvres*, edited by P. Tannery and C. Henry, 1891-4, in 4 volumes. Volume 3 p. 294.

[5] B. Pascal, 'Oeuvres mathématiques', in *Oeuvres complètes*, La Pléiade, 1954, pp. 55-343. See p. 237.

[6] G. Vacca, 'Maurolycus the first discoverer of the principle of mathematical induction', *Bulletin of the American Mathematical Society*, 16 (1970) pp. 70-73.

D. FRANCISCI MAVROLYCI,
ABBATIS MESSANENSIS,
Mathematici celeberrimi,

ARITHMETICORVM LIBRI DVO,

NVNC PRIMVM IN LVCEM EDITI.
Cum rerum omnium notabilium.

INDICE COPIOSISSIMO.

CVM PRIVILEGIO.
Venetijs, Apud Franciscum Franciscium Senensem.
MDLXXV

FIGURE 1 Title page of Maurolico's *Arithmetic*

LIBRI PRIMI
ARITHMETICORVM
MAVROLYCI FINIS.

Completus Messanę in freto Siculo in ædibus ipsius Authoris iuxta Cænobium Carmelitanorum, ad horam noctis secundam diei Dominici, qui fuit Aprilis decimus octauus, & sanctissimum Paschæ festum. Anno salutis.

M. D. LVII.

Maurolico's work is entitled *Arithmeticorum libri duo*, and was printed in Venice in 1575. It is an octavo volume of 198 pages, made up as follows:

- eight unnumbered pages, consisting of the title page, a very full table of figurate numbers, and several pages of general definitions;
- 175 numbered pages, 82 in Book I and 83–175 in Book II;
- the unnumbered *verso* of page 175 is decorated with a printer's vignette which differs from that on the title page;
- two unnumbered pages which form an *Index lucubrationum* constituting a bibliography;
- twelve unnumbered pages entitled *Index coposius in duos libros arithmeticorum* which constitute an alphabetical index of the words and mathematical expressions appearing in the work, with 402 items, which was an exceptional practice at this time.

The first book deals with arithmetic, from the point of view of figurate numbers in the Pythagorean tradition developed by Nicomachus of Gerasa or Boethius; however, Maurolico's presentation will remain without any

166 *Mathematics in the sixteenth century*

doubt as the most comprehensive ever achieved on the subject. The second book deals with the arithmetic of irrational numbers according to Book X of Euclid, but it is original in that it treats this subject as an 'arithmetic'.

In this presentation we shall only deal with the first of these two books of arithmetic.

1. Figurate numbers in Maurolico's first arithmetic Book

The first book of the *Arithmetic* discusses 'figurate' numbers exhaustively. These numbers (the polygonal and polyhedral numbers), which have been discussed since Greek antiquity (Pythagoras, Euclid, Hypsicles, Diophantus), are inductively constructed. It is therefore not surprising to find numerous reasonings by recurrence in this book. But before going any further we shall provide a system of abbreviations in order to simplify the presentation:

Linear numbers
I_n will designate the nth odd number, $2n - 1$, and D_n the nth even number, $2(n - 1)$, because Maurolico begins the sequence of even numbers with 0.

Surface numbers
P_n^k will denote the nth k-gonal number of the first kind, whereas P'^k_n will be the symbol used for the nth k-gonal number of the second kind. (Examples: $P_6^3 = 21$ is the sixth triangular number, P_7^4 is the seventh quadrangular or square number, P'^4_5 is the fifth square number of the second kind, 41.) R_n will designate the nth rectangular number equal to $n(n - 1)$; Maurolico describes these numbers as those 'of which the other part is longer' and thus they are just a special case of 'hetermecic' numbers.[7]

Solid numbers
Δ_n^k and Δ'^k_n will denote pyramidal numbers of the first and second kind respectively. K_n^k and K'^k_n 'column' numbers of the first and second kind. Regular polyhedron numbers which are central or of the second kind will be indicated as Tet'_n, Cub'_n or Hex'_n, Oct'_n, Dod'_n, $Icos'_n$ for, respectively, the nth tetrahedral, cubic or hexahedral, octahedral, dodecahedral, and finally icosahedral numbers.

1.1 *Superficial numbers of the first kind*
The marginal drawings on the first pages of Book I illustrate the constructions of these 'classical' figurate numbers.

---

[7] Maurolico's phrase is '*parte altera longiores*'. D. E. Smith, in his *History of mathematics* (1924, vol. II, pp. 18–19) uses the word 'heteromecic' to translate the Greek *heteromêkis* which the Pythagoreans employed to denote numbers of the form $n(n - 1)$, and the neo-Pythagoreans $n \times p$ with p not equal to n.

Solida Regularia in numeris

1	9	35	91	189	341	559	855	1241	1729	Tetrahedra vel pyramides.
1	15	65	175	369	671	1105	1695	2465	3439	Octahedri et iidem Cubi.
1	33	155	427	909	1061	2743	4215	6137	8569	Icosahedri et iidem Dodecahedri

Quadrati quadratorum

| 1 | 16 | 81 | 256 | 625 | 1296 | 2481 | 4096 | 6561 | 10000 | Biquadrati |

FIGURE 4

consequence of the definitions,[11] where, after having verified the proposition to be proved for the initial index, he stresses the necessity of proving that, if the truth of the proposition is assumed for an arbitrary index, then this entails the truth of the proposition for the next index. For Maurolico the principle is not justified but its use is in most instances correct; he enumerates the numerical cases of verification up to the fifth index, ending with 'and thus repeating to infinity, the proposition is proved', or again 'and in the same manner infinitely, until the proposition is concluded',[12] but Maurolico's use of numerical support goes beyond simple verification because the numbers only serve to illustrate the proof and not to perform it.

Recurrent reasonings begin to appear from the start, for example with proposition 4 (p. 4) which states that odd numbers are obtained by successive additions of the binary to unity. More simply, he starts from the idea that odd numbers are the non-even ones; he has shown (proposition 3) that even numbers are the doubles of roots, in other words that they are obtained starting from 0 by adding 2, then 2, etc. It follows just as $1 + 2 = 3$ that $I_1 + 2 = I_2$, $I_n + 2 = D_n + 1 = 2(n-1) + 1$. Then we find proposition 6 stating that the sum of a root and the preceding one gives the collateral odd number, $I_n = n + (n-1)$; according to proposition 4, $I_n + 2 = I_{n+1}$, thus if $I_n = n + (n-1)$, $I_{n+1} = n + (n-1) + 2 = n + (n+1)$. It is this reasoning which Maurolico uses while relying on $5 + 2 = 7$ and as $5 = 3 + 2$ this gives $3 + (2 + 2) = 3 + 4 = 7$. But the first proposition which is less trivial is proposition 15 which states that the sum of the odd numbers from unity to the nth gives the nth square. To prove this, Maurolico relies on propositions 12 (stating that $n^2 + n + (n+1) = (n+1)^2$) and 13, which follows from the fourth ($n^2 + I_{n+1} = (n+1)^2$); thence, if n^2 is equal to the sum of the odd numbers from 1 to I_n, by adding I_{n+1} the $(n+1)$th square is obtained, which proves the theorem. Maurolico tells us here:[13]

---

[11] B. Pascal, *Oeuvres complètes, op. cit.*, p. 103.
[12] Maurolycus, *op. cit.*. 'et sic deinceps in infinitum semper repetita, propositum demonstratur' and 'eodemque modo infinitum propositio concludit'.
[13] Maurolycus, *op. cit.*, p. 7: 'et sic deinceps in infinitum semper 13^a repetita propositum de monstratur'.

and thus to infinity, by continually repeating the thirteenth [proposition], the proposition proposed is proved.

We also find such classical propositions as proposition 33 (pages 13–14): there are infinitely many integers x, y, z such that $x^2 + y^2 = z^2$; this receives two proofs by recurrence. The first proof shows that if there is an integer k such that

(1) $I_k = (I_n)^2$

then the thirteenth proposition previously cited gives

(2) $I_k + (k - 1^2) = k^2$

It is clear that k is odd [$k = 2n(n - 1) + 1$] and therefore that k^2 is also odd; thus

(3) $I_m = k^2 = [I_{2(n^2 - n + 1) - 1}]^2$

and it is clear that the integer m is larger than k. Since $I_5 = (I_2)^2$, for $9 = 3^2$, there is an infinite sequence of I_k which confirm (1) and therefore confirm (2), which proves proposition 33. For example, starting from $I_5 = (I_2)^2$, $I_5 + 4^2 = I_{13} = 5^2$ is obtained, then $I_{13} + 12^2 = I_{85} = 13^2$ and $I_{85} + 84^2 = 85^2 = I_{3613}$ and so on; thus starting from the Pythagorean triplet (3,4,5) yields an infinite sequence of triplets (5,12,13), (13,84,85), etc.

The second proof which is given is likewise obtained by induction; however, the procedure is a very 'Euclidean' sort which uses segments, so that if one has $x^2 + y^2 = z^2$ then $x^2 \cdot z^2 + (\tfrac{1}{2}y^2)^2 = (x^2 + \tfrac{1}{2}y^2)^2$, since if x is an odd integer then y is even, and $\tfrac{1}{2}y^2$ is again an even integer; starting from the triplet (3,4,5), recurrence gives an infinity of triplets (8,15,17), (32,255,257), (512,65535,65537), . . .

But Maurolico gives proofs by recurrence which are much less simple, in which he exhibits an exceptional virtuosity in the handling of this type of reasoning. For example this occurs in the proof of proposition 90 (pp. 57–59) where it is proved that the triangular sum of the second sort from 1 to the nth odd number is equal to the nth biquadratic gnomon, that is:

$$\sum_{p=1}^{p=I_n} P_p^{'3} = G_n = n^4 - (n - 1)^2$$

Let us observe that the sum which represents the first member of the equality above is by definition the I_nth pyramidal number of the second sort, $\Delta_m^{'3}$. (To simplify the notation, we use m to stand for I_n.) Proposition 79 gives:

(1) $3 \Delta_m^{'3} = K_m^{'3} + P_m^4 + P_m^3$

But by definition $K_m^{'3} = m_m^{'3} = I_n P_m^{'3}$ and $P_m^4 = m^2 = (I_n^2)$, and according to proposition 22, $P_m^3 = n.m = n.I_n$, so that the relation (1) may be written

(2) $3 \Delta_m^{'3} = I_n (P_m^{'3} + I_n + n)$

On the other hand, according to proposition 89,

(3) $3 G_n = 3.I_n.P'^4_n$

which reduces the proof of theorem 90 to that of the equality

(4) $3 P'^4_n = P'^3_m + I_n + n$

Maurolico carries out the proof as follows: he considers the second member of (4) and the difference between two of these expressions for two consecutive values of n. Let us then put $f(n) = P^3_m + I_n + n$, and then form the first and second differences of $f(n)$ for consecutive values of n, that is,

(5) $d_n = f(n + 1) - f(n)$
(6) $e_n = d_{n+1} - d_n$

Then Maurolico shows that e_n is constant and equal to 12 for all n; on the other hand $d_1 = f(2) - f(1) = 15 - 3 = 12$; this implies therefore that $f(n + 1) = f(n) + 12n$. On the other hand, it has been proved by proposition 68 that $P'^4_{n+1} = P'^4_n + 4n$; in other words, $3P'^4_n = 3P'^4_n + 12n$. Since for $n = 1$ one has $f(1) = 3 + 3.P'^4_1$, $f(n) = 3.P'^4_n$ for all n, which had to be proved in order to demonstrate the proposition. It will be realized that reasoning by recurrence is here altogether non-trivial.

Propositions 101 (page 69) and 102 (page 70) may also be cited as statements proved by recurrence. Proposition 101 states that if the sequence of natural numbers is divided into sub-sequences of $1, 3, 5, \ldots, (2n - 1), \ldots$ consecutive terms, the last terms of each sub-sequence form the sequence of squares.

1	2 3	4	5 6 7 8	9	10 11 12 13 14 15	16	.....
1		3		5		7	

Proposition 102 states that if an analogous partition is performed on the sequence of odd numbers, the sums of the sub-sequences thus formed constitute the sequence of biquadratic gnomons:

1	3 5 7	9 11 13 15	17 19 21 23 25 27 29 31 ...
1	15	65	175

$16 = 2^4$

$81 = 3^4$

$256 = 4^4$

2.2 An inductive approach in one proof

It is rare in mathematical works that the procedure which leads to the statement is given; it is known that in arithmetic, induction is an essential heuristic method. Now in the first book of Maurolico's *Arithmetic*, the particulars of the quest for a proof are displayed on pages 62 to 67 with respect to proposition 98. This proposition states that, if G_n designates the nth biquadratic gnomon,

(1) $G_n = (I_n^2) + 2S_n$

where

(2) $S_n = 2P_{n-2}^3 + 3\Delta_{n-1}'^4$

S_n is called the 'supplement' according to ancient tradition. Now it is known from proposition 89 that $G_n = I_n.P_n'^4$. Since, by proposition 68, $P_n'^4 = P_{n-1}^4$, and, from proposition 13, $P_n^4 = P_{n-1}^4$, we have:

$P_n'^4 = 2P_{n-1}^4$ and
$G_n = (I_n)^2 + 2I_n.P_{n-1}^4$

Proving (1) and (2) is thus reduced to proving:

(3) $S_n = I_n.P_{n-1}^4 = 2P_{n-2}^3 + 3\Delta_{n-1}'^4$

With the help of proposition 80: $3\Delta_{n-1}'^4 = K_{n-1}'^4 + 2P_{n-1}^4$, the definition of $K_{n-1}'^4$ and also proposition 68 cited above, we obtain:

(4) $S_n = R_{n-1} + (n-1)\Delta_{n-2}^4 + (n+1)P_{n-1}^4$

So proving (3) is reduced to proving:

(5) $R_{n-1} + (n-1)P_{n-2}^4 = (n-2)P_{n-1}^4$

When Maurolico tries the verification for $n = 2$, this casts no light because the relation is $0 + 0 = 0$. For $n = 3$, this yields $1 \times 2 + 2 \times 1 = 2^2$, but the fact that $n - 2 = 1$ brings about a confusion as to what is to be proved. This no longer occurs for $n = 4$, for which the following schema is given with $p = n - 1$, in other words with $p = 3$:

$pP_{p-1}^4 + R_p = p[P_{p-1}^4 + (p-1)]$
Now, $P_{p-1}^4 + (p-1) = R_p$

according to proposition 55: $pR_p = (p-1) P_p^4$, which gives

(5) $pP_{p-1}^4 + R_p = (p-1)P_p^4$.

Maurolico verifies that this schema is still valid for $n = 5$, that is to say, $p = 4$, and consequently gives the general schema which we have just indicated. Thus the numerical verifications help him to approach the schema which provides the general proof. We are indeed in the presence of a heuristic inductive procedure, not directed towards establishing an assertion but towards obtaining a schema of proof. And this procedure is so exceedingly rare that, in our opinion, attention should be drawn to it.

2.3 Proof of the binomial formula for n = 3,4

The principle of this proof is that of the semi-lattice (*treillis*) of multiples of two numbers a and b, constructed so that the arrow pointing to the left indicates multiplication by a, and the arrow pointing to the right, multiplication by b. Starting from a and b, Maurolico completes the diagram by putting 1 at the apex, and proves in this way a certain number of propositions when $b - a = 1$.

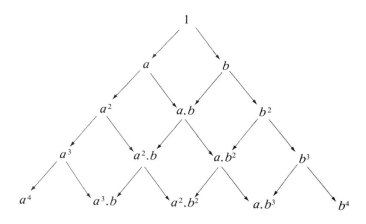

More particularly he proves that the difference between the terms at the extreme ends of one row is equal to the product by $b - a$ of the sum of the elements of the preceding row. He has previously shown that the difference between two adjacent terms of the same row is equal to the product by $b - a$ of the term in the preceding row which generates the two terms in question. Maurolico uses this schema to prove proposition 106 (pp. 72–73).

The first book of Maurolico's *Arithmetic* ends with an appendix, whose proposition 22 is the relation:

$$(a + b)^4 = a^4 + 4a^3.b + 6a^2.b^2 + 4a.b^3 + b^4$$

In this connection, Maurolico says that no Greek or Latin author before him had proved this relation. It may be, then, that he had heard talk of some proofs by the Arabs (al-Karajī for example), since he specifies that he was the first in the tradition of the Greeks and Latins to prove this. The diagrams which illustrate the proof (p. 80), reproduced below, indeed show the path followed and the possibility of extending the proof for $n = 5$, $n = 6$, etc.

178 *Mathematics in the sixteenth century*

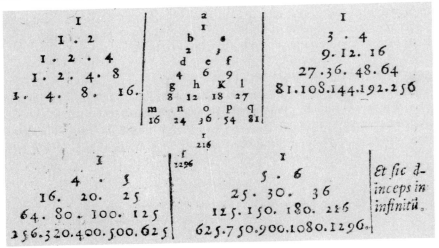

The central schema proceeds as follows:

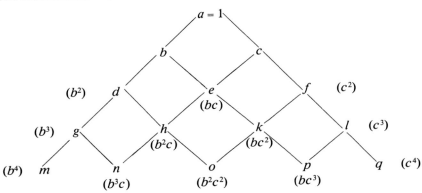

The proof rests on the knowledge of the result $(b+c)^3 = g + 3h + 3k + 1$ (proposition 15 of the appendix). Starting from $(b+c)^4 = (b+c)(b+c)^3$, Maurolico then proves that

$$(b+c)^4 = m + 4n + 6o + 4p + q.$$

One sees that a clever use of the triangle later called Pascal's triangle, together with a general formula for the coefficients, would be sufficient to give a generalization by recurrence of the result obtained here.

2.4 *Sums of finite sequences of numbers in Maurolico's arithmetic*

Without entering into details, a simple enumeration of the sums obtained will suffice to show the richness of the work in this area.

$$\sum_{p=1}^{p=n} P = P_n^3 \quad \text{def}$$

$$\sum_{p=1}^{p=n} P_p^k = \Delta_n^k \quad \text{def}$$

$$\sum_{p=1}^{p=n} P'^k_p = \Delta'^k_n \quad \text{def}$$

$$\sum_{p=1}^{p=n} I_p = P_n^4 = n^2 \quad (\text{prop. 15})$$

$$\sum_{p=1}^{p=n} 2p = R_n$$

$$\sum_{p=1}^{p=n} p^3 = (P_n^3)^2 \quad (\text{prop. 58})$$

$$\sum_{p=1}^{p=n} p'^6_p = \sum_{p=p_{n-1}^3+1}^{p=p_n^3} I_p = n^3 \quad (\text{prop. 42 and 62})$$

$$\sum_{p=1}^{p=n} P_p^4 = \sum_{p=1}^{p=n} p^2 = \Delta_n^4 = \tfrac{1}{3}(K_n^4 + P_n^4 + P_n^3) = \tfrac{1}{3}[n^3 + n^2 + n(n+1)/2]$$

(prop. 63)

$$\sum_{p=1}^{p=I_n} P'^3_p = \sum_{p=(n-1)^2+1}^{p=n^2} I_p = G_n = n^4 - (n-1)^4 \quad (\text{prop. 90 and 101})$$

Within this report it has not been possible to expound all the niceties of proof which the Italian arithmetician carried out. But it is certain that he influenced the French mathematicians of the seventeenth century, such as Pascal, Bachet and Fermat, who were concerned to a greater or lesser degree with arithmetic. Certainly the emergence of the work of Diophantus is a more essential factor in the interest taken in the theory of numbers; but the role played by the publication of Maurolico's *Arithmetic* cannot in any way be neglected.

PART IV

Mathematics and its ramifications

Is translation betrayal in the history of mathematics?

CYNTHIA HAY

1. Is translation betrayal

To give my title in English rather than in the original Italian—*traduttore tradittore*—is, if not a betrayal, certainly a less effective phrasing. Neither the succinctness nor the assonance of the original can be as aptly conveyed in an English version. If the translation of two words has so many drawbacks, the prospects for the translation of a text on a larger scale, whether of poetry or of mathematics, look dismal. Yet the English is not simply a pale and inadequate substitute for the Italian original. There is something to be said in its favour. Translation is always a matter of choices, and the choice of 'betrayal' as a translation elucidates the pithy original. Betrayal connotes breaking faith while appearing to be loyal: it captures a tension basic to translation—translation, for all the considerable efforts that may be made towards fidelity to the original, seems to be doomed to infidelities as well as infelicities.

A sort of schizophrenia sometimes appears to affect assessments of translation. In the past, not just in mathematics, but also in other fields, it has been a major catalyst of intellectual developments. In the history of mathematics, the importance of the translations of key mathematical works in the twelfth and thirteenth centuries, and also in the sixteenth century, is well known, and need not be recounted here. On the other hand, today translation is too often seen as a subsidiary and low-grade activity; translators themselves may suffer from near invisibility, with the credit for their efforts and achievements given in the smallest typeface available at the printers. Yet the translation of any text with a technical content requires considerable specialist knowledge; just as poets are the best translators of poetry, so practising historians are the best qualified to translate both primary and secondary historical material.

For all the length and bulk of words that have been written about translation, there is little, as George Steiner has said, that contributes anything new or original.[1] Much of the second-order discussion of translation is in a disheartening vein; it argues that there are intrinsic difficulties in translation

---

[1] George Steiner, *After Babel: aspects of language and translation*, London, Oxford University Press, 1975, pp. 238-9.

that ensure that any amount of effort and care can only yield at best an inadequate result. The consistent tenor of these warnings may reflect the low esteem in which translation is generally held; by comparison with other intellectual activities, it is seen as something like housework—useful and indeed sometimes unavoidable, but also something that many people prefer to delegate to others.

There are a number of standard categories and contrasts in discussions of translation. Translations are seen as ranging across a spectrum, from the hard case of poetry and the special case of biblical translation to, at the other extreme, a mundane level where translation, in a fairly straightforward way, can achieve both adequacy and accuracy. Corresponding to these different ends of the spectrum are notions of 'high' and 'low' translation. The difficulties of translating poetry are well known, and poets as translators have at times transcended them. But more frequently, translations of poetry have been accompanied by imprecations of betrayal. At the other end of the spectrum, it is argued that simple messages such as those forbidding smoking or the dropping of litter have inter-linguistic equivalents which are both accurate and familiar, so that translation poses no problems. Somewhat lengthier items, conveying information relatively free from cultural peculiarities, are also seen as translatable without serious difficulties. Where translations from the history of mathematics may be placed on this spectrum is problematic; a one-dimensional spectrum to describe the level and the adequacy of translations is itself inadequate.

It has been argued that scientific works can be adequately and accurately translated.[2] Scientific terms are well defined and often international, in the sense that a term coined in one language is then borrowed for other languages; translation is a matter of using the equivalent terms, which are available as a consequence of the nature of scientific development. The claim that the translation of science is straightforward is not one that could now be innocently made. There is a body of work in the sociology of science which has endeavoured to undermine the idea that scientific language is a clean instrument of communication. In this view, science consists of social practices encapsulated and expressed in language; science is seen as the heart of an onion of language. These arguments have the characteristic hubris and failing of sociology; sociologists know better than their subjects what the latter are about. But whether the translation of scientific language is fraught or straightforward does not depend on advocating or disproving a particular line of sociological argument. Detailed investigations of the social life of science and scientists by both historians and sociologists of varied persuasions have given substance to the view that scientific language has rhetorical functions; consequently, difficulties in translating scientific language may

---

[2] e.g. Theodore Savory, *The art of translation*, London, Jonathan Cape, 1957 (new edition 1968), pp. 158–165.

arise. Scientific works may not be simpler to translate.

Translations of scientific and mathematical works from the past may have further difficulties. The historical dimension of many translations does not appear to have been much considered in the literature on translation. Once the translator has been warned that complete success at the appointed task is in principle unattainable, there are a number of practical maxims to hand, for example, literal translation is a low level of endeavour. The translator should strive to transpose the text into something that is not recognizably 'translationese'; that is, to avoid and eliminate translation into 'the third language'[3] located somewhere between the language from and the language to which a text is translated. The notion of a third language is one that strikes a familiar, if uncomfortable note for most translators. Moreover, it is advised that translation of a text from whatever time is inevitably into the language of the time of the translator. This is true in a trivial way. Moreover, self-conscious application of archaisms will hardly improve or historicize a translation. But this maxim does not confront the difficulties of translating documents from the past in an historically sensitive way. If an historian, in giving an account of a topic in the past, tries to convey a sense of its pastness, a translator of an historical document has a similar obligation. One of the ways in which an historian tries to describe and convey the past as past is by indicating how distinctions were drawn differently; and historians as translators do likewise.

In a fascinating paper, Joseph Needham shows how, in old Chinese scientific texts, words have changed their meanings, or, with the passage of time, the same word has referred to related but distinct items of technology:

One of the greatest difficulties about technical terms is that now and again the same word covers two different things; the thing changed while the word remained unchanged. This is a great nuisance from the point of view of the history of technology, because there is nothing for it but to read every possible text you can find in order to throw light on the point when the change in the 'thing' occurred. An outstanding example is [one word] . . . which meant 'copper' long before it meant 'bronze'. So, too, [another] . . . means 'rudder'; but it did also mean 'steering-oar'—and of course one of the most important things in the history of shipping is to know when the steering-oar gave place to the axial or stern-post rudder. In a case like this, all one can do is to read every passage one can find where these words occur, and see what it says.[4]

The notion of 'false friends' in translation takes on a new meaning in the light of Needham's comments. Providing contextually correct translations for key scientific and technical terms is a crucial problem in specialist histori-

---

[3] Alan Duff, *The third language: recurrent problems of translation into English*, Oxford, Pergamon Press, 1981.

[4] Joseph Needham, 'The translation of old Chinese scientific and technical texts', in *Aspects of translation, studies in communication 2*, The Communication Research Centre, University College, London, published by Secker and Warburg, London, 1958, pp. 71–72.

cal translation, and one which can provide historical insights, as Needham suggests.

Barnabas Hughes argues, in connection with his translation of Jordanus de Nemore's *De numeris datis*, that a translation that is faithful in the sense of being literal, to a fault, may be unreadable:

> Two goals were set for the translation and a third was born of necessity. The two were to make it mathematically correct and to render it readable; the third was to give the reader the experience of rhetorical algebra. Rugged as the experience will be, some of the roughness has been eased by removing the rocks of literalness. For example, in I-1 are the sentences '*Sublata ergo differentia de toto, remanebit duplum minoris datum. Qua divisa, erit minor portio data. Sicut et maior.*' Terse as the Latin is, the meaning is clear. The English may be rendered in two ways, (1) literally and (2) liberally.
>
> (1) The difference therefore subtracted from the whole, there will remain given the double of the smaller. Which divided, there will be given the smaller portion. And so the larger
>
> (2) Subtracting therefore the difference from the total, what remains is twice the smaller. Halving this yields the smaller and, consequently, the larger part.
>
> My taste tends to the liberal. Those who prefer a more literal translation may well exercise their skill.[5]

More generally, the degree of liberality appropriate for a particular text depends in part upon the tractability of the original, as well as on questions of the audience or audiences for whom the translation is intended. In the project of translating Chuquet's mathematical manuscript, on the whole, the question of whether Chuquet's sentence structures were to be recast in aid of greater clarity and readability did not arise. It was possible to be fairly literal, in Hughes' sense, and still produce a readable text.

2. Translating Chuquet

The idea of translating Chuquet's mathematical manuscript was due to Graham Flegg. Like many a research idea, this project[6] originated in the context of preparing teaching materials. In developing the materials for the first history of mathematics course at the Open University, AM289, Mr Flegg was both intrigued and puzzled by the various existing assessments of Chuquet's mathematics, and frustrated by the relative inaccessibility of his mathematical works. The whole of Chuquet's mathematical manuscript is in

---

[5] Barnabas B. Hughes, *Jordanus de Nemore: de numeris datis*, Berkeley, University of California Press, 1981.

[6] Three people worked on preparing the translation, Graham Flegg, Barbara Moss and myself. Photographs were made of the manuscripts in the Bibliothèque Nationale, financed by a grant from the Open University Research Committee. The translation was published as G. Flegg, C. Hay and B. Moss (eds.) *Nicolas Chuquet, Renaissance mathematician*, Dordrecht, D. Reidel, 1985.

the Bibliothèque Nationale; the *Triparty* is available in Marre's nineteenth-century edition, and Hervé L'Huillier has published an excellent modern edition of the *Géométrie*.[7] However, Chuquet's mathematical work was not accessible to anyone unable or unwilling to read fifteenth-century French. Although linguistic competence, in as many languages as possible, is an ideal to which we all aspire, not everyone can read all languages with equal ease; English-speaking students who can read languages other than their own are exceptional in more than one sense of the word. Consequently, an English edition of Chuquet's mathematical manuscripts would make his work available to a wider audience. Our original idea was to provide a complete translation of all Chuquet's mathematical manuscripts, with the French original on facing pages. This would have had an estimated length of a thousand pages. The practical realities of publishing in England soon put an end to this idea. Consequently, the translation of Chuquet is intended to be a comprehensive selection from his mathematical manuscripts, with commentary to assist the reader unfamiliar with various aspects of fifteenth century mathematics and its contexts. A selection from Chuquet's manuscripts can provide an introduction to the range of his works; parts of his work are repetitious. For example, there may be little variation in a sequence of problems which Chuquet introduced for pedagogic reasons; another example, from the *Commercial arithmetic*, is his discussion of gold, which follows quite closely his preceding discussion of silver. A scholar who wishes to study, say, the teaching of mathematics in the late fifteenth century, or some aspects of monetary matters, still needs to consult the original for full details. Our edition of Chuquet's manuscripts is intended for readers who might not otherwise encounter primary source material for fifteenth-century mathematics.

One problem in producing a translation of Chuquet was to identify key mathematical terms and phrases and to establish appropriate translations for them. Chuquet is known for his innovations in algebraic notation; his algebraic vocabulary is also well worth consideration.[8] Central to the *Triparty* is the phrase '*rigle des premiers*'. A literal translation of this phrase would be obscure; a free translation, as 'algebra', would be anachronistic. Consequently, the translation used for this phrase, 'rule of first terms', is close to Chuquet's phrasing while making his meaning clear to modern readers.

As Needham has argued, the exercise of translating texts in the history of

---

[7] A. Marre, 'Notice sur Nicolas Chuquet et son *Triparty en la science des nombres*', *Bullettino di bibliografia e di storia della scienze matematiche e fisiche* XIII (1880) pp. 555-659, 693-814, XIV (1881) pp. 413-460; H. L'Huillier, *Nicolas Chuquet, la géométrie*, Paris, Vrin, 1979.

[8] Paul Benoit, 'Recherches sur le vocabulaire des operations elementaires dans les arithmétiques en langue française de la fin du moyen-âge', *Documents pour l'histoire du vocabulaire scientifique* no. 7 (1985), pp. 77-95.

science or mathematics requires careful distinguishing of different senses of words. In translating a substantial body of Chuquet's work, it was clear that he used the word '*partie*' throughout to refer to the *side* of an equation, expressed in varying combinations of words and algebraic notation. Since Chuquet was for the most part a careful and consistent writer, it was a surprise to encounter a passage in his *Géométrie* where he lapsed from this practice and, in the course of one problem, used the word '*partie*' to refer both to the *side* of an equation and to a *part* of a side of a triangle:

Autre stile de investiguer la cathetuse, et ce par la rigle des premiers; et pour ce faire nous prandrons ce triangle qui a 15 et 12 pour ses ypotheneuses et 17 pour la base: ores pour trouver sa cathetuse, il convient de 17 faire deux **parties***, dont l'une multipliee en soy, et celle multiplicacion levee de 15 fois 15, et la reste mise a part, et l'aultre* **partie** *multipliee en soi, et celle multiplicacion levee de 12 foiz 12, la rests soit egale a l'aultre mise a part; et pour ce je pose que la premiere* **partie** *soit 1^1, qui multipliee en soy et soustraicte de 225, restent 225 $\overline{m}\, 1^2$; puis l'aultre* **partie** *qui est $17\,\overline{m}\,1^1$, multipliee en soy monte $289\,\overline{m}\,34^1\,\bar{p}\,1^2$ qui soustraiz de 144, restent $144\,\overline{m}\,289\,\bar{p}\,34^1\,\overline{m}\,1^2$; abrevie tes* **parties***, si auras 34^1 pour partiteur et 370 pour nombre a partir; et par consequent $10\,\frac{15}{17}$ pour quociens, et pour la premiere des deux* **parties***.*

(Another style of investigating the cathetus is by the rule of first terms. To do this we take this triangle which has 15 and 12 for its hypotenuses and 17 for the base. Now to find its cathetus, it is necessary to make two **parts** of 17. One of these multiplied by itself, and this multiplication subtracted from 15 times 15 and the remainder put aside and the other **part** also multiplied by itself and this multiplication taken from 12 times 12, the remainder is equal to the other amount put aside. And for this let us propose that the first **part** be 1^1 which multiplied by itself and subtracted from 225, there remains $225\,\overline{m}\,1^2$. Then the other **part** which is $17\,\overline{m}\,1^1$ multiplied by itself yields $289\,\overline{m}\,34^1\,\bar{p}\,1^2$ which subtracted from 144 there remains $144\,\overline{m}\,289\,\bar{p}\,34^1\,\overline{m}\,1^2$ equal to $225\,\overline{m}\,1^2$. Simplify your **sides** and you will have 34^1 for the divisor and 370 for the number to divide, and consequently $10\,\frac{15}{17}$ for the quotient and for the first of the **parts**.)

The translator is obliged to clarify Chuquet's usage. The exception shows the consistency of Chuquet's practice; it is striking just because he otherwise uses his algebraic vocabulary strictly to treat problems algebraically. Chuquet's lapse on this occasion serves to highlight his achievement in devising and using algebraic terminology in French, an achievement linked with his accomplishments with respect to algebraic notation.

Chuquet's use of algebraic terminology encapsulates an important pattern in the development of algebra. Vernacular mathematics was closely linked with practical and commercial uses. Algebra developed within this framework, as a specialized kind of problem solving in advanced arithmetic. The development of algebraic terminology in the vernacular, in the instance of Chuquet, foreshadowed later developments, namely the gradual movement of algebra away from a kind of practical arithmetic linked with commercial uses, and towards a more general and abstract discipline. The practical and sometimes humble work in algebra in the fourteenth and fifteenth centuries became obscured, and much of it has only recently been recovered by historians.

To return to the considerations discussed earlier in this paper, translation may be a skill too often held in low regard, but it can suggest insights into the history of mathematics. In putting Chuquet into English, the decisions and choices which translation required bring into focus his achievements in algebraic terminology.

Renaissance mathematics (and astronomy) in Baldassarre Boncompagni's *Bullettino di bibliografia e di storia delle scienze matematiche e fisiche* (1868–87)

S. A. JAYAWARDENE

Summary

Some of the most important contributions by nineteenth-century scholars to the historiography of Renaissance mathematics and astronomy are found in the 20 volumes of the *Bullettino* edited and published by Baldassarre Boncompagni. Besides historical studies and articles of bibliographial interest, there are several editions of unpublished texts, among which are the *Regule abaci* of Adelard of Bath, the *Triparty* of Chuquet, and the *Vite di matematici italiani* of Bernardino Baldi. The *Bullettino* is of special interest to codicologists on account of the large number of manuscripts cited in its articles. A cumulative index (of names, subjects, and manuscripts) is in preparation.

Baldassarre Boncompagni (1821–1894) contributed much to our knowledge of Renaissance mathematics and astronomy during the second half of the nineteenth century when the historiography of the physical and mathematical sciences was still in its infancy as an academic discipline. His life and work have been described in the eloquent tributes by his contemporaries and, more recently, in articles in the *Dictionary of scientific biography* and the *Dizionario biografico degli Italiani*.[1]

The unmarried son of a rich Roman patrician family, he lived the life of a recluse, devoting all his time and energy to his activities as a scholar and publisher. Being a bibliophile and a mathematician he was able to build a remarkable library of manuscripts and early printed books of mathematics and astronomy. Two of the periodicals to which he contributed were the

---

[1] Ignazio Galli, *Atti pontif. accad. sci. nuovi lincei*, 47 (1893–94), pp. 161–186; A. Favaro, *Atti. Ist. Veneto Sci.*, (7), 6 (1894–95), pp. 509–521; E. Carruccio, *DSB*, vol. 2 (1970), pp. 283–284; V. Cappelletti, *DBI*, vol. 11, pp. 704–709.

Giornale arcadico di scienze lettere ed arti and the *Atti* of the Accademia de' Nuovi Lincei of which he was a member. In these he published extensive studies of the work of the thirteenth-century astronomers Guido Bonatti and Gherardo Sabbionetta, and of the twelfth-century translators Gherardo of Cremona and Plato of Tivoli.[2] His articles on Leonardo of Pisa published in these journals were followed by separately published editions of the *Liber abaci*, the *Practica geometriae* and *Flos*. He also edited and published the arithmetic of John of Spain and a Latin version of the arithmetic of Al-Khowarizmi.[3] As an ardent bibliophile he amassed a vast amount of bibliographical information on the mathematics and astronomy of the Renaissance, which can be found in his 900-page study of the *Treviso arithmetic* (1478), the first commercial arithmetic to be published in Italy.[4]

In addition to being a scholar and bibliophile, Boncompagni was also the proprietor of a printing press, publishing with the imprint *Tipografia delle scienze matematiche e fisiche*. This press printed and published his own books and also the Transactions (*Atti*) of the Accademia Pontificia de' Nuovi Lincei and the *Bullettino di bibliografia e di storia delle scienze matematiche e fisiche*.

Begun in 1868, the *Bullettino* was edited and published by Boncompagni himself and distributed free to academies, libraries and scholars. Each annual volume on average contained some 500 pages of articles and book reviews and 200 pages of current bibliography. Unfortunately, publication ceased after twenty years because of Boncompagni's ill health and his inability to find a successor. (Antonio Favaro was asked if he would like to take over, but declined as he was already occupied with the editing of Galileo's works.)

The twenty volumes of the *Bullettino* are of interest to us for three main reasons. Many of the articles relate to Renaissance mathematics and astronomy, the subjects dealt with at length being: logarithmic tables, quadrature of the circle, continued fractions, stellated polygons, the *nova* of 1604, mathematics at the College de France, surveyors in the Netherlands, and optical instruments.[5] There are detailed studies devoted to the work of Adelard of Bath, Andalò di Negro, Bartolomeo da Parma, Prosdocimo de Beldomandi, Smeraldo Borghetti, Petrus Peregrinus, Francesco Barozzi and

---

[2] *Giorn. arcad. sci. lett.*, *122* (1851), pp. 138–229; *124* (1851), pp. 186–258; *Atti pontif. accad. sci. nuovi lincei*, *4* (1851), pp. 247–286, 387–493.

[3] *Giorn. arcad. sci. lett.*, *131* (1853), pp. 3–129; *132* (1853), pp. 3–176; *133* (1853), pp. 3–91; *Scritti di Leonardo Pisano*, 2 volumes (Rome, 1857-62); *Atti pontif. accad. sci. Nuovi lincei*, *5* (1851-52), pp. 5–91, 208–246; *Trattati d'aritmetica (I. Algoritmi de numero Indorum, II. Liber algorismi . . . di Johannes Hispalensis)* (Rome, 1857).

[4] 'Intorno ad un trattato d'aritmetica stampato nel 1478', *Atti pontif. accad. sci. nuovi lincei*, *16* (1862-63), pp. 1–64, 101–228, 301–364, 389–452, 503–630, 683–842, 909–1044.

[5] Bierens de Haan *6*, pp. 203–238, *7*, pp. 99–140; Favaro *7*, pp. 451–502, 533–589; Günther *6*, pp. 313–340; Jacoli *7*, pp. 377–405; Sédillot *2*, pp. 343–368, 387–448, 461–510; *3*, pp. 107–170; Vorstermann van Oijen *3*, pp. 323–376; Martin *4*, pp. 165–238.

Zarkali.[6] Lastly, editions of several manuscripts have been published in the *Bullettino*. Among the texts published are:

- several tracts on the abacus, including that of Adelard of Bath[7]
- Bartolomeo da Parma's *Tractatus spherae*[8]
- *Epistola de magnete* of Petrus Peregrinus[9]
- Chuquet's *Triparty*[10]
- Baldi's Lives of the mathematicians[11]
- French translations of the algebraical works of Vieta[12]
- Some mathematical works of Maurolico[13]

The *Bullettino* was not limited to publishing studies of Italian science. In it appeared several studies of science in the Netherlands: Bierens de Haan on the quadrature of the circle and on logarithmic tables; Vostermann van Oijen on Dutch surveyors and their contribution to mathematics.[14] Most important of all was Bierens de Haan's 400-page bibliography of historic works of mathematics and physics published in the Netherlands.[15]

There is a book-length essay by Sédillot on professors of mathematics and physics at the College de France.[16] He has also contributed articles on Islamic science. Other writers on Islamic science were Steinschneider and Hankel. Steinscheider's main contribution was a study of Zarkali, eleventh-century Arab astronomer.[17] Hankel's article was an Italian translation of the chapter on Arab mathematics in his *Zur Geschichte der Mathematik in Altertum und Mittelalter* (1874).[18]

The main contributor to the journal, apart from Boncompagni, was Antonio Favaro, whose articles, if collected together, would have filled two volumes. The greater part of his writings, however, relates to the life, work, and times of Galileo Galilei—in particular, to unpublished manuscripts and

[6] Boncompagni *14*, pp. 1–90; Simoni *7*, pp. 313–338; Boncompagni *7*, pp. 339–376; Narducci *17*, pp. 1–42; Favaro *12*, pp. 1–74, 115–251; *18*, pp. 405–423; Boncompagni *13*, pp. 1–80, 121–200, 245–368; Bertelli *1*, pp. 1–32, 65–99; Boncompagni *17*, pp. 795–848; Steinschneider *14*, pp. 171–182; *16*, pp. 493–504; *17*, pp. 343–360; *18*, pp. 343–360; *20*, pp. 1–36, 575–604.

[7] Treutlein *10*, pp. 589–594; Gerland *10*, pp. 595–647; Boncompagni *10*, pp. 648–656; *14*, pp. 91–134.

[8] *17*, pp. 43–120, 165–218.

[9] Bertelli *1*, pp. 65–99.

[10] *13*, pp. 593–659, 693–814.

[11] *5*, pp. 427–534 (notes by Steinschneider); *7*, p. 337; *12*, pp. 420–428; *19*, pp. 335–406, 437–489, 521–640 (notes by Narducci); *20*, pp. 197–308 (notes by Narducci).

[12] *1*, pp. 223–276.

[13] *9*, pp. 23–121.

[14] See note 5.

[15] *14*, pp. 519–630, 677–717; *15*, pp. 225–312, 355–440; *16*, pp. 393–444, 687–718.

[16] See note 5.

[17] See note 6.

[18] *5*, pp. 343–401.

archive material in the libraries of Florence. Of these writings, one article of interest to us is his reconstruction of the catalogue of Galileo's library.[19] As for mathematics before 1600, we have historical notes on continued fractions, and a study of the fifteenth-century Paduan mathematician Prosdocimo de Beldomandi.[20]

Favaro has also written several essay reviews among which are those of: Wolf's *Geschichte der Astronomie* (1877), Moritz Cantor's *Vorlesungen über Geschichte der Mathematik* (1880), Prowe's *Copernicus* (1883-4), and Eugen of Sicily's translation of Ptolemy's *Optics*.[21] Other essay reviews are: by Moritz Cantor of Kuckuck's arithmetic in the sixteenth century and of a new edition of the *Cartelli* exchanged between Ferrari and Tartaglia[22]; by Mansion of Hankel's history of mathematics in antiquity and the Middle Ages; by Hoüel of Friedlein's history of number symbols and arithmetic.[23]

Among the manuscripts owned by Boncompagni were a collection of 202 biographies of mathematicians written at the end of the sixteenth century by Bernardino Baldi of Urbino. Many of these were published in the *Bullettino*. Fourteen biographies of Arab mathematicians were edited and annotated by Steinschneider. Another fifty-seven, of Italian mathematicians, were annotated by Enrico Narducci (1832–1893), Boncompagni's secretary and Librarian of the Biblioteca Alessandrina.[24] Narducci also published in the *Bullettino* the annotated edition of the *Tractatus spherae* of Bartolomeo da Parma and Baldi's life of Pythagoras. Other articles by him are a study of a fourteenth-century Italian translation of Alhazen's *Optics* and a bio-bibliography of the historian of mathematics, Franz Woepcke (1826–1864).[25]

Boncompagni's own contributions to the *Bullettino*, besides 167 editorial notes, consisted of 87 articles running to some 1200 pages. Among these were: a study of Albiruni's book on India; a catalogue of the works of Andalò di Negro; notes on Widmann's arithmetic; a lengthy introduction to the text of Adelard's *Regule abaci*; detailed notes to Baldi's lives of John of Saxony, John Lineriis and Luca Pacioli, and to the arithmetic of Smeralda Borghetti of Lucca.[26] Besides writing the obituary of Michel Chasles, Boncompagni provided the obituary notices of Hermann Hankel, Gottfried Friedlein and Louis Amélie Sédillot with bibliographies of their works.[27]

The *Bullettino* is of special interest to the codicologist as its pages contain

---

[19] *19*, pp. 219-293; *20*, pp. 372-376.
[20] See note 5 and note 6.
[21] *11*, pp. 757-777; *14*, pp. 183-205; *16*, pp. 333-348; *18*, pp. 327-331; *19*, pp. 115-120.
[22] *9*, pp. 183-187; *11*, pp. 177-196.
[23] *8*, pp. 185-220; *3*, pp. 67-90.
[24] See note 11.
[25] *17*, pp. 1-120, 165-218; *20*, pp. 197-308; *4*, pp. 1-48; *2*, pp. 119-152.
[26] *2*, pp. 153-206; *7*, pp. 339-376; *9*, pp. 188-210; *14*, pp. 1-90; *12*, pp. 352-438; *13*, pp. 1-80, 121-200, 245-368.
[27] *13*, pp. 815-827; *9*, pp. 297-308; *9*, pp. 536-553; *9*, pp. 656-700.

194 *Mathematics and its ramifications*

references to some 2500 manuscripts in about 60 cities. This figure does not include documents in archives and Galilean manuscripts. Besides editions of manuscripts (already referred to) there are also inventories of manuscripts of Andalò di Negro, Bartolomeo da Parma, Bernelinus, Gerland, Ibn al-Haytham, Petrus Peregrinus, and Zarkali.[28]

To exploit the wealth of material in the *Bullettino* detailed indexes are necessary. Each volume is already provided with an index of names—in which the location of every name appearing in an article, footnote or bibliography has been given. In the last volume, there are:

- a list of articles, volume by volume, in the order in which they appear
- an index of authors and titles of articles
- an index of hitherto unpublished documents[29]

To complement these indexes I am compiling a cumulative index of names, subjects and manuscripts edited. It is also proposed to cumulate (using a computer) the indexes of names in the twenty volumes.

---

[28] *7*, pp. 339–376; *17*, pp. 16–31; *14*, pp. 75–79; *10*, pp. 648–656; *14*, pp. 721–740; *1*, pp. 67–70; *16*, pp. 493–496.

[29] *20*, pp. 697–712, 713–728, 729–749.

Bhaskara problem is similar to the Brahmagupta problem in having four pipes, but it has different rates.[11]

Sanford (1927, p. 70) says a version appears in the *Propositiones ad acuendos iuvenes*[12] which is attributed to Alcuin. (The earliest known partial version is ninth-century, but the first complete version is late tenth century and some of the problems may have been added up to then.) However, the only possible candidate is problem 8 of Alcuin, which reads as follows:

A cask and three pipes.

A cask is filled to 100 *metretae* capacity through three pipes. One-third of its capacity plus 6 *modii* flows in through one pipe; one-third of its capacity flows in through another pipe; but only one-sixth of its capacity flows in through the third pipe. How many *sextarii* flow in through each pipe?[13]

Clearly this is not a cistern problem in the usual sense, and Folkerts[14] notes this, though Tropfke[15] does mention it as a cistern problem.

A proper version appears in Mahavira (*c.* 850)[16] and then numerous variations occur in Fibonacci (including a lion, leopard and bear eating a sheep),[17] the Columbia Algorism, Paolo dell'Abbaco, Chuquet, Borghi, Widmann, Calandri, Tartaglia, Buteo, etc., on down to the modern version 'The Pope can pray a soul out of purgatory in one day, a cardinal in three days, . . .'

This evidence from Heron and the *Greek anthology* suggests Greek origins for the problem. (It is unclear whether the *Greek anthology* was known to Alcuin, Fibonacci, etc. The tenth century edition was compiled by Constantinus Cephalas at Constantinople. It was much rearranged by Maximus Planudes, who was Byzantine Ambassador to Venice in 1327. Planudes' version displaced Cephalas' and a unique manuscript based on Cephalas was rediscovered in the Palatine Library at Heidelberg in 1606/7 (MacKail, 1890).)

However, two other sources are now known.

First, the *Chiu Chang suan shu* is now available in a German edition by K. Vogel.[18] This appears to have been compiled about 150 BC, with the existing version dating from the first or second century AD. Chapter VI, problem 26[19] is a proper cistern problem with five pipes. Problems 22, 23 and 25 are

---

[11] *ibid.* p. 42.
[12] Alcuin (1863); Folkerts (1978).
[13] From the answer and other problems, a *metreta* is 72 *sextarii* and a *modius* is 200 *sextarii*. I am indebted to John Hadley for translating Alcuin for me.
[14] Folkerts (1978) p. 36.
[15] Tropfke (1980) pp. 578-579.
[16] Mahavira (1912) chapter 8, v. 32-34, p. 266.
[17] Fibonacci (1857) pp. 182-186.
[18] See the bibliography entry headed *Chiu Chang Suan Shu* (Vogel 1968).
[19] *op. cit.*, pp. 68-69.

related problems where one is given the rates for two or three processes and one wants to carry out the same number of each process in a day.

Second, the *Bakhshali* manuscript contains versions with three and with seven rates of giving money. The rates are complexly expressed, e.g. $\frac{7}{2}$ dinars in $\frac{4}{3}$ days.[20] Unfortunately, controversies over the date of this manuscript make it difficult to reach firm judgements as to its role in the transmission of problems. If we accept Hoernle's late nineteenth-century arguments,[21] then it is *c.* 400; however, Tropfke (1980) places it *c.* 1000. Even more unfortunately, there does not appear to be any complete translation of the text.

This problem already indicates many of the features which I have noted in looking for the sources of such problems. These are familiar to historians if not to mathematicians. First, the sources are hard to find—one needs access to several major libraries and even then success is not assured. Secondly, references are wrong or inaccurate often enough that one must actually examine the original material as far as possible. As just seen, two of the first-millenium sources of the cistern problem cited in competent histories turned out to be non-existent, and the references to Heron and Brahmagupta were inaccurate.

In a project of this scope, I am not able to examine all the original manuscripts, so I am dependent upon facsimiles, printed texts and careful translations of them, such as the work of Arrighi, Folkerts, Hunger, Lam, Libbrecht, Vogel, etc., who have recently greatly extended the available texts. It has been noted already at this conference that this has its hazards, so my bibliography always gives a careful citation for each reference. It is also true that each new text may radically change the history of a problem. Indeed, I had not seen Tropfke or the *Chiu Chang Suan Shu* before the conference and they have radically changed the history of the cistern problem. Tropfke also cites the problem in several Arabic sources, filling in the missing transition from Mahavira to Fibonacci. The Indians now seem as likely to have obtained it from China as from Greece.

In my 196-odd sections, I have only nine problems which I conclude to be of Greek origin:[22]

> The loculus of Archimedes.
>
> 'Euclid's' ass and mule problem.
>
> Men find a purse.
>
> Men buying a horse.
>
> Aristotle's wheel paradox.
>
> The paradox of the liar.

[20] Kaye (1927) pp. 49–52.
[21] Hoernle (1888).
[22] For the second, third and fourth problems on this list, see Tropfke (1980) pp. 606–611.

The isoperimetric problem.

Archimedes' cattle problem.

Casting out nines.

(I am presently adding Archimedes' determination of the volume of two intersecting cylinders.)

Tropfke[23] notes that simple forms of 'Euclid's' problem occur in the *Chiu Chang Suan Shu*, though they seem closer to 'men buying a horse' to me.[24] He also cites Chang Ch'iu-chien (*c.* 475), without indicating just what form of the problem occurs. He does not note that Diophantus gives a general solution (I, 15) and even considers three- and four-person variants (I, 18 and 19).[25] Tropfke also does not note that Mahavira gives variants with three people.[26]

Similarly Tropfke (1980, p. 608) shows that the *Chiu Chang Suan Shu* has some problems similar to 'men buying a horse', but does not note that Diophantus (I, 24 and 25)[27] has general solutions for three and four people. So, though there may still be some doubts, I think these two problems still belong to the Greek tradition. Nonetheless, it is clear that little recreational material can be considered as being of Greek origin.

It may be worth noting that the Roman tradition contributes almost nothing at all. I only have Ovid's game, the posthumous twins inheritance problem,[28] and possibly 'the problem of the pandects' where two men with differing amounts of food share each with a third man, though my earliest source for this is Fibonacci.[29]

Now let me describe some problems whose origin is fairly well established as being Oriental.

Knight's tours. Mathematicians tend to associate these with Montmort and de Moivre (early eighteenth century).[30] Lucas[31] cites Paolo Guarino di Forli (1512). But the histories of chess by van der Linde (1874) and Wieber (1972) give references back to the dawn of the game, e.g. an Arabic manuscript of 1141 by Abu Ishaq[32] gives these tours.

Monkey and coconuts. This is a modern formulation and may not be well known to you, so I give it here:

Five sailors and their pet monkey are shipwrecked on an island. They spend all day gathering a pile of coconuts and decide to divide them in the morning. In the night,

---

[23] *ibid.* p. 610.
[24] *Chiu Chang suan shu*, ed. Vogel (1968), book VIII, problems 10, 12, 13, pp. 86–88.
[25] See Heath (1964) pp. 134–136.
[26] Mahavira (1912) chapter 6, v. 251–258.5, pp. 158–159.
[27] Heath (1964) pp. 139–140.
[28] Tropfke (1980) pp. 655–658.
[29] Fibonacci (1857) p. 283.
[30] See for example Ball (1974) p. 176.
[31] Lucas (1882–94), volume 4, p. 207.
[32] Wieber (1972) pp. 479–480.

one sailor awakes and decides to take his fifth. He divides the pile into five equal parts, but there is one coconut extra which he gives to the monkey. He then takes and hides his fifth and puts the rest back in a pile. Then another wakes up and does exactly the same, and then a third and a fourth and a fifth. In the morning, they divide the pile into five equal parts, again finding one extra coconut which they give to the monkey. How many coconuts were there?

This problem was used in a short story called 'Coconuts' by the American writer Ben Ames Williams in *The Saturday evening post* in 1926. It produced an enormous flood of letters and telephone calls, amusingly described by Martin Gardner,[33] who says the problem goes back to 1912. However, both Sam Loyd and H. E. Dudeney give complex versions, but with the final remainder given, which much reduces the difficulty and interest of the problem. In looking through Renaissance material, I discovered that the problem was well known then. Fibonacci gives several forms of it, but always in determinate form, i.e. with the final remainder specified. These usually involve a merchant paying fees at various cities or a man stealing or carrying apples who has to bribe or pay several porters. Mahavira gives several versions and I was pleased to find that he has an indeterminate one.[34] As usual, Tropfke[35] has references to sources that I have not seen. Determinate versions are in the *Chiu Chang Suan Shu*[36] and the *Bakhshali* manuscript.[37] These often specify the total gain or loss rather than the remainder. There is also a simple determinate version in the Papyrus of Akhmin (c. seventh century) described in Tropfke (1980) and in Heath's *History*.[38]

The hundred fowls problem. Tropfke[39] and I are in reasonable agreement over the earliest sources. However, there is a major spanner in the chronology, namely the *Bakhshali* manuscript, which I have mentioned above. If its date is c. 400, then this manuscript has the first appearance of the problem,[40] preceding Chang Ch'iu-Chien (c. 475), which is the generally accepted source.[41] There were a number of Chinese commentators on Chang, which are described in Libbrecht (1979) and Mikami (1913)[42] but are omitted in Tropfke, probably because they are so nonsensical. Mahavira[43] has a general technique and several examples, including the earliest where the number of birds and the amount of money are not the same. Alcuin has eight

---

[33] Gardner (1958) pp. 118 ff.
[34] Mahavira (1912) chapter 6, v. 131.5, pp. 124–125.
[35] Tropfke (1980) pp. 582–586.
[36] *Chiu Chang Suan Shu* book VI, problems 27–28, pp. 69–70; book VII, problem 20, pp. 79–80.
[37] Hoernle (1888) p. 277; Kaye (1912) p. 358.
[38] Heath (1981) volume 2, p. 544.
[39] Tropfke (1980) pp. 572–573 and 613–616.
[40] Bag (1979) p. 92; Kaye (1927) p. 42.
[41] Libbrecht (1979) p. 277; Mikami (1913) p. 43; Needham (1959) p. 122.
[42] Libbrecht (1979) p. 278–282; Mikami (1913) pp. 43–44.
[43] Mahavira (1912) chapter 6, v. 146–153, pp. 132–135.

Some early sources in recreational mathematics 203

versions,[44] including his no. 39 about 'An Oriental merchant' who is buying camels, asses and sheep. I find the appearance of these problems in Alcuin quite mysterious as Alcuin nominally precedes any transmission to the West via the Arabs and some of these problems are present in the earliest version (ninth century), though no. 39 first appears in the next oldest manuscript of c. 1000. No. 39 is numerically identical with the first problem of abu Kamil (c. 900).[45] If the first manuscript of Alcuin is actually a bit later, then Abu Kamil could be the route of transmission from the Orient to Europe, especially as we know no other early Arabic writer on this problem. On the other hand, abu Kamil's treatment is quite elaborate—one problem has five kinds of birds, leading to two equations in five unknowns with 2676 positive integral solutions which he claims to have found. His introductory comments may be referring to other authors' attempts at such problems and this, together with the detail of his treatment, may well indicate that there were earlier Arabic works on the problem which could have been known to Alcuin. (Tropfke also cites Tabari, who is late eleventh-century Persian, and al-Kashi, who is fourteenth or fifteenth century in Samarkand, so neither seems a likely link.) After Alcuin, there are dozens of examples in the West. Chuquet[46] gives a version and gives solutions (3, 3, 24) and (4, $\frac{2}{3}$, 25$\frac{1}{3}$). The latter is permissible because cloth is being purchased. Chuquet says one can find as many solutions as one wants!

A number of other classical problems are clearly of Oriental origin:

Sharing an unequally distributed cost, e.g. the cost of stairs in a multiply occupied house or the cost of a partly dug well, occurs in Mahavira and then in dell'Abbaco.[47]

The chessboard problem.[48]

Chinese rings.[49]

Magic squares.[50]

Chinese remainder theorem.[51]

Selling different amounts at the same prices and getting the same. Here Tropfke[52] begins with Fibonacci, but Mahavira[53] has six examples which

---

[44] Nos. 5, 32, 33, 33a, 34, 38, 39, 47 in Alcuin (1863) and Folkerts (1978). See Folkerts (1978) p. 37.
[45] Suter (1910-11); Folkerts (1978) p. 37.
[46] Chuquet (1881) problem 83; see pp. 213-214 in the 1984 edition.
[47] Tropfke (1980) pp. 528-530.
[48] ibid. pp. 630-633.
[49] Afriat (1982); Needham (1959) p. 111.
[50] Needham (1959) pp. 55-62.
[51] Tropfke (1980) pp. 636-642.
[52] ibid. pp. 651-652.
[53] Mahavira (1912) chapter 6, v. 102-110, pp. 113-116.

204 *Mathematics and its ramifications*

are so complex as to be incomprehensible and Bhaskara II [54] has an example which is still complex but is comprehensible.

Present of gems.[55]

The three sisters.[56]

Overtaking and meeting problems (e.g. hound and hare).[57]

Snail climbing out of well.[58]

(Tangrams[59] seem also to be Oriental, but they are only known back to the early nineteenth century, with a variant back to the mid-eighteenth century.)

A frequent characteristic of these is their appearance in various Chinese and Indian sources, especially Mahavira, and then in Western material such as Alcuin or Fibonacci, without obvious connecting links of Arabic material. Tropfke does help out here in citing some Arabic and Persian texts and translations, but almost all are too late to influence Alcuin and almost all are unavailable in European languages.

From Alcuin on, a number of new problems arose in the West. Alcuin alone contains five new problems among its fifty-three problems:

River crossing problems (three types no less!)[60]

The transport problem, later called the jeep or explorers' problem, where explorers or transport carry provisions, but also consume provisions as they go, and one wants to get across a desert. Alcuin's problem[61] concerns camels carrying grain. His answer is confusing, but Folkerts makes it clear. My only other pre-twentieth century reference for this problem is Luca Pacioli,[62] a reference kindly provided by Professor Franci. Martin Gardner[63] discusses modern versions. Folkerts does not mention Pacioli or the modern versions and says the problem never reappears.

Division of casks and contents.[64]

Combining prices and amounts incorrectly, often called the apple-seller's problem.[65]

Strange families, e.g. widow and son marrying widow and son.[66] Folkerts[67]

[54] Colebrooke (1817) pp. 242–244.
[55] Mahavira (1912) chapter 6, v. 162–166, pp. 137–138.
[56] Tropfke (1980) p. 642.
[57] *ibid.* pp. 588–598.
[58] *ibid.* pp. 591–598.
[59] Needham (1959) pp. 111–112.
[60] Alcuin (1863) and Folkerts (1978), nos. 17–20; Tropfke (1980) pp. 658–659.
[61] Alcuin (1863) and Folkerts (1978), no. 52; see Folkerts p. 38.
[62] Agostini (1924), problems 49–52, p. 156.
[63] Gardner (1961), problem 1.
[64] Alcuin (1863) and Folkerts (1978), no. 12; Tropfke (1980) pp. 659–660.
[65] Alcuin (1863) and Folkerts (1978), no. 6; Tropfke (1980) p. 652.
[66] Alcuin (1863) and Folkerts (1978), no. 11. The Bedeversion gives two more examples (11a and 11b in Folkerts 1978).
[67] *op. cit.* p. 38.

says these seem to be of Roman origin but he gives no references.

At the conference, I said that the remarkable amount of new material in Alcuin and the lack of any clear linkage to the Orient was one of the most interesting problems raised by this project and I find it remains a tantalizing question.

Some other problems arising during the Middle Ages and Renaissance are the following:

The Josephus problem.[68]

Fibonacci's rabbits, which number about 3.2×10^{1963}, as of the end of 1984.

Fibonacci appears to be the first to give an algorithm for Egyptian (i.e. unit) fractions.

Inheritance problem where the ith gets $i + \frac{1}{a}$ of the remainder. (Graham Flegg pointed out that Chuquet[69] gives versions of this with non-integral numbers of heirs!)[70]

Jug problem—divide 8 in half using jugs of size 5 and 3.[71]

After about 1250, there seems to be a hiatus in new problems until Dürer, Cardan and Tartaglia, and this corresponds to the fact that there was a period of consolidation after the twelfth-century renaissance.

Conclusions

The history of mathematical recreations is large, but it provides a microcosm of the history of mathematics and of culture as a whole. I have found that tracing particular problems gives a clearer view of the passage of knowledge from a culture to its successors.

As a mathematician, rather than an historian, I found that the Oriental sources were much more numerous and more interesting than I had expected. There is no doubt that recent work, much of it inspired by Needham (1959), has transformed our understanding of Chinese mathematics. So far, few histories of mathematics have taken proper account of Oriental work. (I only know of Tropfke (1980) and of van der Waerden's recent, somewhat idiosyncratic, book (1983).) Hence the average mathematician is quite unaware of the extent of the Oriental contribution. I was particularly impressed by reading Mahavira, whose problems often seem as modern as Tartaglia's or even Martin Gardner's.

At the conference, I said that it appeared that there must have been more and earlier transmission from the Orient via the Arabs than we are now

---

[68] Ahrens (1910–1918) pp. 118–169; Smith (1929) vol. II, pp. 541–544; Tropfke (1980) pp. 652–654.
[69] Chuquet (1881), pp. 224–225 in 1984 edition.
[70] Tropfke (1980) pp. 586–588.
[71] ibid. p. 659.

aware of and this has been noted by others (Vogel, 1968; Folkerts, 1978). Tropfke (1980) provides some references in this gap but many of them are not yet available in Western languages.

As indicated at the beginning, this is a report on work in progress, so that the conclusions are tentative and some of them may be well known already to historians. There is still a great deal of material to be examined. Nonetheless, I hope this preliminary report shows that the history of mathematical recreations is an interesting and amusing part of the history of mathematics.

Bibliography

Afriat, S. (1982). *The ring of linked rings*. Duckworth, London, 1982.

Agostini, A. (1924). 'Il "De viribus quantitatis" di Luca Pacioli', *Periodico di matematiche* (4), 4 pp. 165-192.

Ahrens, W. (1902). ' "Nim", ein amerikanisches Spiel mit mathematischer Theorie', *Naturwissenschaftliche Wochenschrift* 17:22 (2 March 1902), pp. 258-260.

Ahrens, W. (1910, 1918). *Mathematische Unterhaltungen und Spiele*. Second edition, 2 volumes, Teubner, Leibzig.

Alcuin (1863). *Propositiones alcuini doctoris Caroli Magni Imperatoris ad acuendos juvenes*. In: *Patrologia latina*, ed. J.-P. Migne, tom. 101, Paris, 1863, columns 1143-1160. (A different version is in the works of Bede, *ibid.*, tom. 90, Paris, 1904, columns 667-672. See also Folkerts, *infra*.)

Bag, A.K. (1979). *Mathematics in ancient and medieval India*. Chaukhambra Orientalia, Varanasi, Delhi.

Ball, W.W. Rouse (1974). *Mathematical recreations and essays*. 12th edition revised by H.S.M. Coxeter, University of Toronto Press. (The first edition was published in 1892, Cambridge University Press.)

Biggs, N.L., Lloyd, E.K. and Wilson, R.J. (1976). *Graph theory 1736-1936*. Oxford University Press, 1976.

Bouton, C.L. (1901/2). 'Nim: a game with a complete mathematical strategy', *Annals of mathematics*, (2) 3, pp. 35-39.

Chiu Chang suan shu (1968). (Nine chapters on the mathematical art.) Translated by K. Vogel as *Neun Bücher arithmetischer Technik*, Vieweg, Braunschweig.

Chuquet, N. (1881). 'Problèmes numériques faisant suite et servant d'application au Triparty en la science des nombres de Nicolas Chuquet Parisien'. Edited and published by A. Marre as 'Appendice au Triparty en la science des nombres de Nicolas Chuquet parisien', *Bull. bibliog. storia sci. mat. fis.* 14, pp. 413-460. Page references are to the partial translation in Flegg, G., Hay, C. and Moss, B. eds., *Nicolas Chuquet, Renaissance mathematician*, Reidel, Dordrecht, 1984 (copyright date 1985).

Colebrooke, H.T. (1817). *Algebra, with arithmetic and mensuration from*

Some early sources in recreational mathematics   207

the Sanscrit of Brahmegupta and Bháscara. John Murray, London, 1817; reprinted by Sändig, Wiesbaden, 1973.

Datta, B. and Singh, A. N. (1962). *History of Hindu mathematics*. Parts I (1935) and II (1938) in a combined edition. Asia Publishing House, Bombay.

Dickson, L. E. (1952). *History of the theory of numbers*. 3 volumes, reprinted by Chelsea, New York.

Fibonacci, L. (1857). *Liber abaci*. Volume I of *Scritti di Leonardo Pisano*, ed. by B. Boncompagni, Rome.

Folkerts, M. (1978). 'Die älteste mathematische Aufgabensammlung in lateinischer Sprache: Die Alkuin zugeschriebenen *Propositiones ad acuendos iuvenes*'. Österreichische Akademie der Wissenschaften, math.-naturwiss. Kl., Denkschriften 116:6, pp. 14–80. (Also printed separately by Springer, Vienna, 1978.)

Gardner, M. (1958). 'The monkey and the coconuts', *Scientific American* 198:4 (April 1958), pp. 118ff. Reprinted in *The 2nd scientific American book of mathematical puzzles and diversions*, Simon and Schuster, New York, 1961, pp. 104–111. (UK edition entitled: *More mathematical puzzles and diversions*, Penguin, 1966, pp. 82–87.) The problem and solution first appeared in May and June 1959, but the book contains additional material.

Gardner, M. (1961). 'Nine more problems'. Chapter 14 of *The 2nd scientific American book of mathematical puzzles and diversions, supra*.

Heath, T. L. (1964). *Diophantus of Alexandria*. Second edition, corrected. Dover, New York.

Heath, T. L. (1981). *A history of Greek mathematics*. 2 volumes, corrected reprint, Dover, New York.

Heron (1914). *Peri metron*. In *Heronis Alexandrini opera quae supersunt omnia*, edited J. Heiberg, vol. V, p. 176–177, problems 20 and 21. Teubner, Leipzig. Reprinted 1976.

Hoernle, A. F. R. (1888). 'The Bakhshali manuscript', *Indian antiquary* 17, pp. 33–48 and 275–279.

Kaye, G. R. (1927–33). 'The Bakhshālī manuscript—a study in medieval mathematics', *Archaeological survey of India—new imperial series* 43:1-3. (All my references are to Part 1.)

Kaye, G. R. (1912). 'The Bakhshālī manuscript', *J. Asiatic society of Bengal* (2) 8, pp. 349–361.

Libbrecht, U. (1973). *Chinese mathematics in the thirteenth century*. MIT Press.

Lucas, E. (1882–94). *Récréations mathématiques*. 4 volumes, Gauthier-Villars, Paris. (There were second editions of volumes 1 and 2.) Reprinted, using first edition of volume 2, Blanchard, Paris, 1975–77.

MacKail, J. W. (1890). *Select epigrams from the Greek anthology*. Longmans, Green and Co., London. pp. 12–26 and 295.

Mahāvīrā (or Mahāvīrācārya) (*c. 850*). *Ganita-sāra-sangraha (or -samgraha)*. Translated by M. Rangācārya. Government Press, Madras, 1912.

Metrodorus (1916–1918). 'Arithmetical problems', in *The Greek anthology*, translated by W. R. Paton. Loeb Classical Library, Putnam's, New York and Heinemann, London. (All the mathematical material is in Book 14, which is in volume 5 of this edition.)

Midonick, H., ed. (1968). *The treasury of mathematics*. Revised edition. 2 volumes. Penguin.

Mikami, Y. (1913). *The development of mathematics in China and Japan*. Teubner, Leipzig. (Reprinted Chelsea, New York, 1961?)

Needham, J. (1959). *Science and civilization in China. volume III— Mathematics and the sciences of the heavens and the earth*. Cambridge University Press, Cambridge.

Sanford, V. (1927). *The History and significance of certain standard problems in Algebra*. Teachers College, Columbia University, New York, *Contributions to Education* no. 251; reprint: AMS Press, New York, 1972.

Sanford, V. (1930). *A short history of mathematics*. Houghton Mifflin, Boston. Reprinted 1958.

Schaaf, W. L. (1970–78). *Bibliography of recreational mathematics*. 4 volumes. National Council of Teachers of Mathematics, Reston, Virginia. (My volume 1 is the 4th edition, 1970. I think that this was the last edition and that the other volumes have only one edition.)

Smith, D. E. (1923). *History of mathematics*. 2 volumes. Ginn, New York. Reprint: Dover, New York, 1958.

Smith, D. E. ed. (1929). *A source book in mathematics*. 2 volumes. Reprint: Dover, New York, 1959.

Struik, D. J. ed. (1969). *A source book in mathematics 1200–1800*. Harvard University Press, 1969.

Suter, H. (1910–11). 'Das Buch der Seltenheiten der Rechenkunst von Abū Kāmil el Misrī', *Bibliotheca mathematica* (3) 11, pp. 100–120.

Tropfke, J., revised by K. Vogel, K. Reich and H. Gericke (1980). *Geschichte der Elementarmathematik. I. Arithmetik und Algebra*. Fourth edition, completely rewritten. De Gruyter, Berlin.

van der Linde, A. (1874). *Geschichte und Literatur des Schachspiels*. 2 volumes. Springer, Berlin. (Reprint in one volume: Olms, Zurich, 1981.)

van der Waerden, B. L. (1983) *Geometry and algebra in ancient civilizations*. Springer, Berlin.

Wieber, R. (1972). *Das Schachspiel in der arabischen Literatur von den Anfängen bis zur zweiten Hälfte des 16. Jahrhunderts*. Verlag für Orientkünde Dr. H. Vorndran, Walldorf-Hessen.

Williams, B. A. (1926). 'Coconuts', *The Saturday Evening Post*, 9 October 1926, pp. 10–11, 186, 188.

Cornelius Agrippa's mathematical magic

A. GEORGE MOLLAND

THIS is a disreputable subject, and demands an apologia, if not an apology. The first part of this is easy, for in recent years great, and sometimes exaggerated, emphasis has been placed on the role of magic in the emergence of modern science. It is hence readily seen to be a worthy object of scholarly study, but it is not so clear that this tolerance should immediately be extended to its mathematics, for this has tended to embarrass even those writers most sympathetic to the magical tradition.[1] Nevertheless it was enthusiastically accepted by many intelligent men of the Renaissance, and this in itself demands empathy and explanation. But, even so, is Agrippa a satisfactory choice as its representative? He was certainly not the most original or systematic thinker of his era, and indeed is someone easily accused of charlatanry. However, these very features, together with the wide diffusion of his works, help to ensure that he accurately reflects the spirit of the age. By the same token, we shall not need to linger over the problem of sincerity posed by the fact that his magical 'encyclopedia' *De occulta philosophia* was published only shortly after his *De incertitudine et vanitate scientiarum et artium* in which he had roundly attacked magic.[2] A first, and much shorter, version of the former work had been completed in 1510, but in this article I shall only concern myself with its final form of 1533.[3]

1. Devices

The *De occulta philosophia* is divided into three books, dealing respectively with natural, mathematical and ceremonial magic, which in turn correspond

---

[1] See e.g. F. Yates, *Giordano Bruno and the hermetic tradition* (London 1964; repr. New York, 1969), p. 324.
[2] Cf. C.G. Nauert, *Agrippa and the crisis of Renaissance thought* (Urbana, 1965), pp. 32-33; D.P. Walker, *Spiritual and demonic magic from Ficino to Campanella* (London, 1958), pp. 90-91; Yates, *op. cit.* (note 1), p. 131.
[3] A facsimile of the 1510 manuscript is published in Agrippa, *De occulta philosophia*, ed. K.A. Nowotny (Graz, 1967). My references will be to Agrippa, *Opera* (Lyon [?], n.d.; repr. Hildesheim, 1970). On the printing history of *De occulta philosophia* and other matters of Agrippan bibliography see J. Ferguson, 'Bibliographical notes on the treatises *De occulta philosophia* and *De incertitudine et vanitate scientiarum* of Cornelius Agrippa', *Publications of the Edinburgh bibliographical society*, 12 (1925), pp. 1-23, and Nauert, *op. cit.* (note 2) pp. 32-33, 106, 112-113, 335-338.

(albeit somewhat loosely) to the three-fold division of the world into elemental, celestial and intellectual. Book II begins with a panegyric on mathematics:

> The mathematical disciples are so necessary and cognate to magic that, if anyone should profess the latter without the former, he would wander totally from the path and attain the least desired result. For whatever things are or are effected in the inferior natural virtues are all effected and governed by number, weight, measure, harmony, motion and light, and have their root and foundation in these.[4]

Agrippa's first examples of the power of mathematics do not at first glance appear particularly magical. They concern the making of automata, of things which, although lacking natural virtues, are yet similar to natural things. Tradition had handed down quite a copious list of these, and Agrippa cites examples attributed to Vulcan, Daedalus, Archytas, and Boethius.[5] Of similar kind were simulacra produced by optical means in the manner taught by Apollonius and Witelo, and Agrippa claims for himself the knowledge of how to make something that can seem suggestive of the telescope:

> And I have come to know how to make two alternate mirrors in which, when the sun is shining, everything that is illuminated by its rays is clearly discerned at remote distances of several miles.[6]

It was knowledge of mathematics and natural philosophy that led to the construction of all these, and the same had been true for greater marvels in Antiquity, such as the Caspian Gates and the Pillars of Hercules, of which now only vestiges or reports remained:

> Although all these seem to be in conflict with nature itself, yet we read that they were made, and to this day we discern their traces. The crowd say that suchlike were the work of demons, since the arts and the artificers have perished from memory, and there are not those who bother to understand and examine.[7]

So far we do not seem to be straying from the history of technology as ordinarily understood, but, after brief references to magnetism, Agrippa makes it clear that there is more to his mathematics than to ours:

> And here it is appropriate for you to know that just as by natural things we acquire natural virtues, so by abstract, mathematical and celestial things we receive celestial virtues, namely motion, life, sense, speech, foreknowledge and divination, even in less well disposed matter, such as that fashioned not by nature but only by art. And in

---

[4] *De occulta philosophia* II.1 (*Opera*, I, p. 153); cf. *De incertitudine et vanitate scientiarum et artium* XLI (*Op.*, II, p. 89).

[5] *De occ. phil.* II.1 (*Op.*, I, pp. 153-154). Cf.*De inc. et van.* XLIII (*Op.*, II, pp. 91-92), and J.P. Zetterberg, *'Mathematicall magick' in England: 1550-1650* (Ph.D. thesis, University of Wisconsin, 1976), pp. 34-35.

[6] *De occ. phil.* II.1 (*Op.* I, p. 154). Cf. *De inc. et van.* XXVI *(Op.* II, pp. 60-61).

[7] *De occ. phil.* II, 1 (*Op.* I, p. 155).

this way images that speak and predict the future are said to be made, as William of Paris narrates about the speaking head forged at Saturn's rising, which they say spoke with human voices.[8] He who knows how to choose suitable matter that is best fitted to be the patient, and also the strongest agent will produce indubitably more powerful effects. For it is a general axiom of the Pythagoreans that just as mathematicals are more formal than physicals so they are more actual, and just as they are less dependent in their being so also are they in their operation, and among all mathematicals, as numbers are more formal so also are they more actual; to them are attributed virtue and efficacy for both good and evil, not only by the philosophers of the heathens but by the theologians of the Hebrews and Christians.[9]

With this emphasis on the power of number Agrippa has reached the heartland of his mathematical magic, and can leave behind his consideration of mechanical and optical devices, whose construction was subjected to geometry rather than to arithmetic.

2. Numbers

At this point it is necessary to emphasize as strongly as possible that Agrippa's conception of number was very different from our own. For him, as for the preceding philosophical and pure mathematical tradition, a number was a collection of units. In this conception fractions and surds were not properly numbers, for true numbers were very definitely discrete entities, to be contrasted with the continuous quantities that were the concern of geometry. Agrippa was conscious that other views were possible, but insisted that his concern was with

rational and formal number, not the material, sensible or vocal number of the merchants, about which the Pythagoreans, the Academics and our Augustine care nothing.[10]

It is only when this is clear that we can hope to make sense of Boethius' assertion (highlighted by Agrippa) that numbers provided the exemplar for the Creation, or realize how significant was Kepler's break with this numerological tradition in asserting that, on the contrary, geometry was God's archetype.[11]

In talking of formal numbers Agrippa and his like were asserting the existence of entities that were both outwith the soul and distinct from the

---

[8] Cf. L. Thorndike, *History of magic and experimental science* (New York, 1923-58), II, p. 351.
[9] *De occ. phil.* II.1 (*Op.* I, pp. 155-6).
[10] *De occ. phil.* II.2 (*Op.* I, p. 157).
[11] Boethius, *De institutione arithmetica* I.2, ed. G. Friedlein (Leipzig, 1857; repr. Frankfurt, 1966), p. 12; cf. Agrippa, *De occ. phil.* II.2 (*Op.* I, p. 156). C.G. Jung and W. Pauli, *The Interpretation of nature and the psyche* (London, 1955), pp. 159-167; J.V. Field, 'Kepler's Rejection of numerology', *Occult and scientific mentalities in the Renaissance*, ed. B. Vickers (Cambridge, 1984), pp. 273-296.

numbered objects, whereas opponents of magic such as Descartes would be equally insistent that numbers were no different from the things being counted.[12] Although separate, formal numbers were yet manifested in all sorts of collections in the world, and it is here that the modern reader has a particular difficulty in rendering numerology intelligible, for the question immediately comes to mind of where a particular collection, say a flock of sheep of which some members are beginning to stray, ends. In one sense this difficulty is by no means new, and Plotinus, for example, emphasized that not every group of ten men that you counted formed a true decad, but only something like a choir or an army.[13] However, in the present time we are faced with the additional problem that our conception of the physical world does not allow numerous finite and well-defined sets of things of the same kind. In Agrippa's day the situation was very different, and it is enlightening to remember how even somewhat later Kepler struggled to provide a geometrical rather than a numerological answer to the question of why there were precisely six planets.[14] The reason we do not ask this question is not so much that we think that Kepler had the wrong number, but that we think that the number is basically accidental, whereas for Kepler it was embedded in God's original design for the universe.

Agrippa devotes several chapters of his *De occulta philosophia* to discussing particular manifestations of the smaller numbers at various levels of his hierarchical universe. Many of the results are presented in tabular form, and we may conveniently read down the scale of Two (*binarius*).[15] In the archetypal world this is represented by names of God of two letters each (in Hebrew): *Iah* (from jod and hev) and *El* (from aleph and lamed). In the intellectual world we have two intelligible substances, angel and soul, and in the celestial world two great luminaries, Sun and Moon. In the elemental world two elements, namely earth and water, give rise to the living soul, and the lesser world has two principal seats of the soul, heart and brain. Finally, when we reach the infernal world, we find that there are two leaders of the demons, Behemoth and Leviathan, and two things which Christ threatens to the damned, namely weeping and gnashing of teeth. Quotation of part of the surrounding text will make it clear that this list includes only some of the mysteries of Twoness.

Plutarch writes that the Pythagoreans called unity Apollo, the dyad strife and boldness, the triad justice, which is the highest perfection, but nor is this without many mysteries. Hence two tables of law in Sinai, two cherubims facing the mercy

---

[12] R. Descartes, *Regule ad directionem ingenii* XIV, in *Oeuvres de Descartes*, ed. C. Adam and P. Tannery (Paris, 1897–1913), X, pp. 445–446.
[13] *Enneads* VI.6.16.
[14] Cf. Field, *op. cit.* (note 11).
[15] I use capitals to distinguish the nouns *binarius*, *ternarius*, etc. from the adjectives *duo*, *tres*, etc.

seat with Moses, two olives pouring out oil in Zecharia,[16] two natures in Christ, divine and human. Hence two appearances of God were seen by Moses, front and rear.[17] Also two testaments, two commands of charity, two first dignities, two first peoples, two genera of demons, good and bad, two intellectual creatures, angel and soul, two great luminaries, two equinoxes, two poles, two elements giving rise to the living soul, earth and water.[18]

As usual the reasoning connecting together the different manifestations of the number, and relating these to any of its mathematical properties is rather loose, and even Agrippa's meaning is sometimes obscure.

We may complement our vertical tour of the manifestations of Two by a horizontal trip along the manifestations of the first twelve numbers at the celestial level.[19] We find that there is one king of the stars, two great luminaries, three quaternions of signs and of houses and three lords of triplicities, four triplicities of signs, four groups of stars and planets related to the elements and four qualities of celestial elements, five erratic stars, six planets deviating from the ecliptic in the breadth of the zodiac, seven planets, eight visible heavens, nine mobile spheres, ten spheres of the world, twelve signs of the zodiac and twelve months. Eleven does not appear on the list, because:

just as it transgresses ten, which is of law and commands, so it falls short of twelve, which is of grace and perfection. It is therefore the number of sinners and penitents. Hence in the tabernacle there were ordered to be made eleven blankets of hair-cloth,[20] which is the dress of penitents and bewailers of their sins. Wherefore this number has no communion with divine things, nor with celestial ones.[21]

In his discussion of the celestial manifestations of number, Agrippa makes quite a lot of capital from a relatively small investment. Much of the discussion depends either directly or indirectly on their being just seven planets with certain differentiae between them. The other main theme is the division of the zodiac into twelve signs. This was somewhat less natural, but relatively few in Agrippa's time would have regarded it as purely conventional in the way that Nicole Oresme had suggested in the fourteenth century, although it was standard to hold that the division was in need of justification.[22]

The hierarchy and numerology in Agrippa's universe are combined with an elaborate system of relations and correspondences between its elements. These may be on one level, as when:

---

[16] *Exodus* XXXI.18, XXV.20; *Zechariah* IV.12.
[17] *Exodus* XXXIII.11, 22–23.
[18] *De occ. phil.* II.5 (*Op.* I, pp. 162–163).
[19] *De occ. phil.* II.4–14 (*Op.* I, pp. 161–200).
[20] *Exodus* XXXVI.14–17.
[21] *De occ. phil.* II.14 (*Op.* I, p. 196).
[22] S. Caroti, 'Nicole Oresme, Quaestio contra divinatores horoscopios', *Archives d'histoire doctrinale et littéraire du moyen age*, *43* (1976), pp. 201–310, at pp. 251–252. Cf. *Supplementum ficinianum* ed. P. O. Kristeller (Florence, 1937; repr. 1973), II, pp. 41–42.

Mercury, Jupiter, Sun and Moon are friends to Saturn, and Mars and Venus his enemies. All the planets are friends of Jupiter except Mars, and so all except Venus hate Mars. Jupiter and Venus love Sun, while Mars, Mercury and Moon are hostile. All except Saturn love Venus. Friends of Mercury are Jupiter, Venus and Saturn, enemies Sun, Moon and Mars. Jupiter, Venus and Saturn are friends of Moon, and Mars and Mercury enemies.[23]

Alternatively there may be correspondences between different levels and kinds of thing. The most famous of these are probably those that subordinate different parts of the body to different signs of the zodiac and also to different planets, but the system also extends to elements, metals, precious stones, animals, plants, etc., etc..[24] Moreover, the relationships are mirrored at the various different levels:

> And of what sorts are the friendships and enmities of the superiors so also are the inclinations of the things subordinated to them among the inferiors[25]

We are thus drawn irresistibly towards the image of a series of structure-preserving mappings.

At this point there is the temptation to go further and speak more generally in terms of structuralism or of abstract algebra. As regards the former, there do seem to be definite affinities between Agrippa's thought and structuralism, but I am not sure how illuminating it would be to describe him as a proto-structuralist or to use modern structuralist techniques to analyse his system. The question of abstract algebra is related but tighter, and an attempt has been made to portray geomancy, a divinatory technique on which Agrippa himself wrote,[26] in group-theoretic terms.[27] However, the group involved was rather trivial, and in general there is something absurd in trying to assimilate Agrippa and his like, as regards aims, methods and attitude, to modern algebraists. But with all this said, there does seem to be a certain similarity of mental *set* involved, which may perhaps be explicated by the counterfactual speculation that, if it had been Renaissance magic that developed into modern science, the appropriate mathematics would have been more akin to abstract algebra than to the infinitesimal calculus. It would have been a mathematics that concentrated on discreteness rather than on continuity.

3. Magic squares

But let us return to Agrippa, and this time to a rather more complex example of his numerology, namely the role of magic squares.

[23] *De occ. phil.* I.17 (*Op.* I, p. 33).
[24] *De occ. phil.* I.22-32 (*Op.* I, pp. 43-59).
[25] *De occ. phil.* I.17 (*Op.* I, p. 34).
[26] *In geomanticam disciplinam lectura* (*Op.* I, pp. 500-526).
[27] M. Pedrazzi, 'Le figure della geomanzia: un gruppo finito abeliano', *Physis*, *14* (1972), pp. 146-161.

Cornelius Agrippa's mathematical magic 215

> There are also handed down by the mages certain mensules of numbers, distributed among the seven planets, which they call the holy tables of the planets, distinguished with very many and great virtues of celestials, inasmuch as they represent the divine rationale of celestial numbers, impressed on celestials by the ideas of the divine mind through the mediation (*ratio*) of the soul of the world and the sweetest harmony of the celestial rays, according to the ratio of effigies that consignify the supramundane intelligences, which can only be expressed by the signs of numbers and characters, for material numbers and figures only have power in the mysteries of hidden things representatively through formal numbers and figures, inasmuch as these are governed and informed by divine intelligences and numerations . . .[28]

The 'mensules' are in fact magic squares, of which one is associated with each of the seven planets: a 3 by 3 one for Saturn, 4 by 4 for Jupiter, and so on, up to 9 by 9 for the Moon.[29] In close conjunction with this each planet also has three characteristic diagrams.

Let us look at the example of Saturn. Agrippa describes the situation thus:

> The first of these 'mensules', assigned to Saturn, consists of a square Three, containing nine individual numbers and in each line three, which, in whatever direction and along each diameter, constitute fifteen; the total sum of the numbers is forty-five. Over this are set by divine motions names that fill up the aforesaid numbers, with an intelligence for good and a demon for evil. From the same number is elicited the sign or character of Saturn and his spirits, which we shall annex to his table below. They say that this table engraved on a lead plate with Saturn favourable aids birth, makes a man secure and potent, and brings about the success of petitions to princes and powers, but, if it is made with Saturn unfavourable, it hinders building and planting and suchlike, dispossesses a man of honours and dignities, engenders strife and discord, and scatters armies.[30]

Associated with Saturn are various names, whose numerical values in Hebrew characters correspond to the magic square, being either 3, 9, 15 or 45.[31] For example, the intelligence of Saturnis called Agiel, which Agrippa forms from the Hebrew characters yaleph, gimel, jod, aleph, lamed, and accordingly has numerical value $1 + 3 + 10 + 1 + 30 = 45$. Similarly the name of the demon, Zazel, is formed from zain, aleph, zain, lamed, and has value $7 + 1 + 7 + 30 = 45$.

Saturn himself and his two spirits also have characteristic diagrams associated with them (see Fig. 1), which are in some way derived from the magic square. K. A. Nowotny has proposed an elaborate explanation of this

---

[28] *De occ. phil.* II.22 (*Op.* I, pp. 215–216).
[29] On ways of associating magic squares with the planets see W. Ahrens, 'Magische Quadrate und Planetenamulette', *Naturwissenschaftliche Wochenschrift*, N.F. *19* (1920), pp. 465–475. See also I. R. F. Calder, 'A note on magic squares in the philosophy of Agrippa of Nettesheim', *Journal of the Warburg and Courtauld Institutes*, *12* (1949), pp. 196–199, and M. Folkerts, 'Zur Frühgeschichte der magischen Quadrate in Westeuropa', *Sudhoffs archiv*, *65* (1981), pp. 313–338.
[30] *De occ. phil.* II.22 (*Op.* I, p. 216).
[31] *De occ. phil.* II.22 (*Op.* I, p. 219).

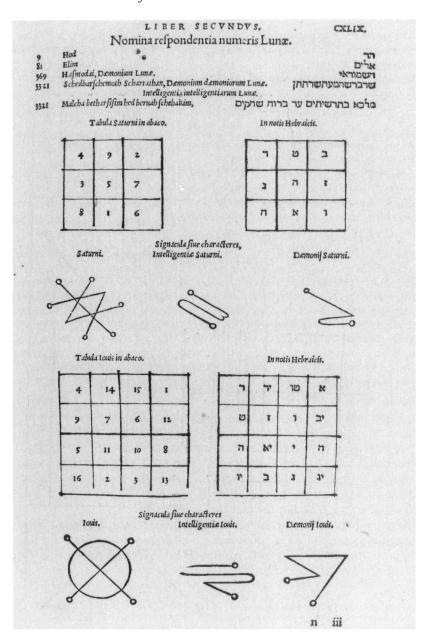

FIGURE 1. H. C. Agrippa, *De occulta philosophia, Liber secundus*, CXLIX, 1510. By courtesy of the British Museum.

derivation,[32] which I do not find convincing in all its details. Nevertheless it does seem clear from his work that the principal rationale for the formation of the characters of Saturn's two spirits (and also the spirits of the other planets) is to join the numbers in the magic square that correspond to the Hebrew letters of the spirit's name. (A number greater than nine may be reduced to one of one digit by adding together its digits, which is equivalent to division by nine and considering only the remainder.) Thus the geometrical figures have an arithmetical origin, which enhances their power, for, as Agrippa later remarks, 'Geometrical figures arising from numbers are thought to be of no less powers.'[33]

Agrippa should not be thought of as deploying any purely mathematical originality in his treatment of magic squares; in this respect his work is almost certainly completely derivative.[34] His concern is more with using the magic squares and associated names, numbers and figures to provide a set of meaningful denominations for the planets. When Alice tentatively asked Humpty Dumpty whether a name had to mean something, she received the testy answer: 'Of course it must: my name means the shape I am—and a good handsome shape it is, too. With a name like yours, you might be any shape, almost'. Again, with the computerization of libraries, we are being asked to focus on bar codes, names which possess only denotation, as well as on shelf marks, which, in systems such as Dewey, are also connotative. Agrippa was firmly in the tradition which believed that things had their correct names, which reflected their true natures and were thus possessed of considerable power. He may without much exaggeration be thought of as putting forward a scientific programme in which the properties of things would be derived from such names, although the programme of course remained vague, and had little hope of being significantly realized.

4. Occult virtues

Finally, instead of following Agrippa through all the other topics of Book II, I wish to turn to the question of occult virtues.[35] These played a key role, at least rhetorically, in the transition to modern science, and can provide a test case for a programme of mathematization. They were traditionally defined as properties of an object that did not derive in any direct way from the elements that constituted it. There was dispute over their causation, but quite frequently they were assigned irreducibly to a body's specific substantial

---

[32] K.A. Nowotny, 'The construction of certain seals and characters in the work of Agrippa of Nettesheim', *Journal of the Warburg and Courtauld Institutes*, 12 (1949), pp. 46-57.

[33] *De occ. phil.* II.23 (*Op.* I, p. 216).

[34] I am grateful for Dr. J. Sesiano's opinion on this question.

[35] Cf. K. Hutchison, 'What happened to occult qualities in the scientific revolution?', *Isis*, 78 (1982), pp. 233-253, for a somewhat different view of these.

218   *Mathematics and its ramifications*

form, and thus had the character of brute inexplicable facts. In the fourteenth century Nicole Oresme had directed his doctrine of configurations of qualities to the issue. In this doctrine causal efficiency was assigned to the pattern of variations of intensities of qualities across a subject, and Oresme suggested that by means of it

> there could be briefly assigned a general rationale of certain occult virtues and marvellous effects or experiments whose causes are otherwise unknown.[36]

The programme is to make the virtues mathematically intelligible. In similar but stronger vein, Descartes held it as a major virtue of his mathematico-mechanistic programme that it showed the way to finding the causes of the 'wonderful effects which are usually referred to occult qualities',[37] and accusations of resorting to occult qualities soon became a standard form of abuse from mechanically inclined philosophers.

Agrippa had a considerable interest in occult virtues, and in a passage that appears heavily dependent on Albertus Magnus enumerates opinions on their causation.

> Alexander the Peripatetic, not deviating from his qualities and his senses, thinks that they arise from elements and in fact from their qualities, which could perhaps be thought true, except that the qualities are of the same species, but many operations of stones agree neither in genus nor in species. Therefore the Academics, together with their Plato, attribute the virtues to the formative ideas of things, but Avicenna reduces such operations to intelligences, Hermes to stars, and Albert to the specific forms of things.[38]

Unlike Albertus, Agrippa held that, rightly understood, none of these opinions deviated from the truth.

> In the first place, God, the end and origin of all virtues, presents a sigil of ideas to his ministers the intelligences, who, like faithful executors, consign, like instruments, whatever is entrusted to them to the heavens and stars, which are meantime disposing the matter to receive those forms which, as Plato says in the *Timaeus*, reside in the divine majesty through the stars, that are to be drawn out. The giver of forms distributes them through the ministry of the intelligences, which he sets over his works as mistresses and custodians, to whom there is entrusted in the things committed to them such a faculty that every virtue of stones, herbs, metals and all the rest is from these presiding intelligences. Thus form and virtue arise first from the ideas, then from the guiding and governing intelligences, afterwards from the disposing aspects of the heavens, next from the disposed complexions of the elements, corre-

---

[36] *Nicole Oresme and the medieval geometry of qualities and motions. A treatise on the uniformity and difformity of intensities known as Tractatus de configurationibus qualitatum et motuum*, ed. M. Clagett (Madison, 1968), p. 236.
[37] R. Descartes, *Principia philosophiae* IV, 187, in *ed. cit.* (note 12), VIII-1, p. 314.
[38] *De occ. phil.* I.13 (*Op.* I, p. 26). Cf. Albertus Magnus, *Book of minerals*, II.i.1–4, tr. D. Wyckoff (Oxford, 1967), pp. 55–67.

score (2,1) and a DS [1 + (4x − 1); 2 − 4x], we go to (2,2) with a DS of [1;1]. According to the stated principle, the second player has to win (2 − 4x) and the first obviously will lose (4x − 1). These two amounts have to be equal. Adding as said in the text 1 ducat to either side of the equation

$$4x - 1 = 2 - 4x$$

we get

$$4x = 3 - 4x.$$

The rest of the text gives the solution $x = \frac{3}{8}$ and the admonition to proceed in the same way in similar problems. The solution says that at a score of (2,0) the first player is entitled to $2x$ or $\frac{3}{4}$ of his opponent's ducat, or to $\frac{7}{8}$ of the whole stake. This is correct, if one assumes that both players have an equal chance of winning a single game, because in this case the second player in order to win the stake has to win the next three games in a row. This will happen only in one case out of eight. Moreover the principle underlying the solution is generally valid if chances of winning a single game are equal for both players. One can see that immediately:

Let the expectation of the first player at a score *(m,n)* be *E(m,n)*. Then $E(m,n) - \frac{1}{2} E(m + 1,n) + \frac{1}{2} E(m,n + 1)$ holds. This again is equivalent to

$$E(m,n) - E(m,n + 1) = E(m + 1,n) - E(m,n).$$

E(m,n) − E(m,n + 1) is either the loss of the first player in case he loses the *(m + n + 1)*th game or the winnings of the second player. Accordingly *E(m + 1,n) − E(m,n)* is the winnings of the first player in case he wins the *(m + n + 1)*th game. Thus the principle is valid and it allows us to see immediately what still seemed to be curious to Pascal,[26] namely that the winnings of a player from winning the first game and from winning the second game are equal. This principle suffices to solve other cases of the problem of stakes, as is indicated in the last sentence of the text concerning this special case.

The text continues with a similar problem of points at a score (3,0) when four wins are necessary for winning the stake. Unfortunately, already the second step in the calculation makes it clear that the author did not stick to this principle and so went astray, with the result that he breaks off after a few lines.

This situation opens the way to differing interpretations. Either we have two authors dealing with two instances of the problem of points or only one. In either case it is clear that the author who tried the problem with four games to win had no idea what this principle meant. If he had really understood the solution of the problem with three games to win, he could never have failed in the more complicated case in the way he did. In the light of the later

---

[26] Letter to Fermat of July 29, 1654; *Oeuvres de Fermat*, vol. II, p. 294.

discussions of Luca Pacioli in the Venetian domain as well as of the Tuscan Calandri[27] at the end of the fifteenth century, one must conclude that this astounding and correct solution had no aftermath. Moreover as the manuscript testifies, somebody who knew it did not understand it. One possible explanation for this, which I utter with all caution, is that we are here confronted with a piece of mathematical work that had been developed in an intellectual environment different from fourteenth-century Tuscany, and at any rate far away from the mathematical style of Calandri, Pacioli, and his sixteenth-century critics. A possible locus for such an intellectual environment, at least for a short period, is the Islamic world, in which some terms concerning games of chance still used in the Western world had come into being. Certainly no Arabic manuscript containing the mathematical solution of a gaming problem could have survived the more or less periodic outbreaks of religious fanaticism and the accompanying purification campaigns in Islamic history. So this early solution of the problem of points as well as the treatment of a dicing problem in the poem *De vetula* from the mid-thirteenth century could be testimonies to what had once been known to mathematicians in Islam.[28] The oldest known printed source for the treatment of the problem of points is Luca Pacioli's *Summa* of arithmetic and geometry.[29]

Pacioli's prominence in the mathematical literature is undisputed. In a citation index of the sixteenth century he would figure very prominently, even if most of the references to his name started with *error di Fra Luca*. At least in contrast to many others he had managed to find a printer for his huge mathematical collection. This was also the reason why everybody who dealt with the problem of points in the sixteenth century referred exclusively to Luca Pacioli. Since Pacioli himself refers to older opinions on how to solve the problem of stakes which he judged to be preposterous, and since Tartaglia and Forestani accepted these opinions as they can bere constructed from Pacioli's account, Luca Pacioli had a very strong impact on the development of the problem of stakes. It is interesting for us that Pacioli looks on his own solution as the 'truth and the right way'.[30] Pacioli was still convinced of the existence of a uniquely determined solution to the problem. The model for this solution is the *compagnia* or *societas* in which the profits and losses are shared according to the amounts invested by the members. Pacioli demonstrates that he was inextricably bound to his own economic model when he rejects an opinion which differs from his own:

[27] See Laura Toti Rigatelli, 'Il "Problema delle parti" in manoscritti de XIV e XV secolo', *Mathemata* (ed. M. Folkerts and U. Lindgren), Franz Steiner, Wiesbaden and Stuttgart, 1985 (= Boethius vol. 12), pp. 229–236.

[28] Compare Ivo Schneider, 'Luca Pacioli und das Teilungsproblem: Hintergrund und Lösungsversuch', *Mathemata* (ed. M. Folkerts and U. Lindgren), Franz Steiner, Wiesbaden and Stuttgart, 1985 (= Boethius vol. 12), pp. 237–246.

[29] Luca Pacioli, *Summa de arithmetica geometria proportioni et proportionalita*, Venice, 1494 and Toscolano, 1523.

[30] 1494 edition, fol. 197^r.

Do not proceed like some others who refer to the game of *morra* and say, if one in a game with five fingers has 4 and the other 3, 'let us go back by one' so that one has 2 and the other 3. For this is not fair, because one resigns $\frac{1}{3}$ and the other $\frac{1}{4}$ of his claim, so that they don't resign to the same extent.[31]

One understands from this quotation that it was not the logical inconsistencies of this 'reduction method' that drove Pacioli to reject it, but only its incompatibility with his own solution, which prescribes sharing the stakes in proportion to the number of games won. Only to those who accept Pacioli's way of sharing the stakes as reasonable does this refutation appear convincing. Those who published were not only convinced that Pacioli was wrong but also that the solutions he proposed were debatable. The first to question them was Girolamo Cardano in his *Practica arithmetica* (Milano, 1539). Typical of the change that had taken place in the half century between Pacioli and Cardano is that now the main point is the subjective acceptability of a solution. For Cardano Pacioli's solution is totally unacceptable because it leads to absurdities from examples. In one of these he presents a case where the stakes should go to the player who first wins nineteen games and the game is interrupted at a score of 18:9. According to Pacioli, the first player is entitled to $\frac{2}{3}$ and the second to $\frac{1}{3}$ of the stakes to which each player had contributed twelve gold coins. Cardano then argues:

The winner of eighteen games will have gained from his opponent only four gold coins, which is $\frac{1}{3}$ of his stake, and yet he lacks only one game to obtain the total winnings, whereas the second lacks ten. This is totally absurd.[32]

Now we should bear in mind that Cardano had already offered his own solution to the problem of points.[33] This solution prescribes that if one player needs r and the other s games in order to win the stakes, these have to be divided in the ratio

$$^{s+1}C_2 : {}^{r+1}C_2.$$

In this case we have $r = 1$ and $s = 10$. Accordingly the stakes should be divided in the ratio of

$$^{10+1}C_2 : {}^{1+1}C_2 = 55:1.$$

But instead of referring to this solution Cardano continues in the following way:

Furthermore, either one may take a part which he could reasonably stake under the same conditions. Thus he who had won eighteen could stake 10:1 even 20:1 against somebody who has nine in a game for nineteen. Therefore he is entitled to twenty parts and the other to one only.

[31] *Ibid.*, fol. 197ᵛ.
[32] Cardano, *Practica* . . . , 1539, fol. 289ᵛ.
[33] *Ibid.*, fol. 143.

Here the subjectivity of the solution becomes perfectly clear. The leading player is entitled to a part of the stakes which is fifty-five times the share of the other according to Cardano's own solution. But Cardano prefers to estimate a factor and starts with ten in order to end with twenty. Already with this procedure it is obvious that Cardano has given up the idea of a unique solution to the problem. But even a Cardano who would have stuck to his own solution would appear to have a very different way of thinking from that of Pacioli. Let us look therefore at Cardano's own solution. Cardano had proposed that if the match ends when one of the players has won n games and if the first player has won r games and the second s games—in symbols, *(r,s)*—then they should share the stakes in the ratio of the *progressiones* of *(n − s)* and *(n − r)*, that is to say:

$$^{n-s+1}C_2 : {}^{n-r+1}C_2.$$

Cardano presents a 'demonstration' of this solution which is based on the special case $r = n - 1, n - s = i, i = 1,2,3$. This 'demonstration' has been interpreted by Coumet in a very convincing way.[34]

1. To begin with, let us suppose that Primus and Secundus each stake a ducat and that they play according to this convention that we shall call c_1, that whoever first wins a game wins the contest, and obtains the adversary's ducat.

2. Let us now set the problem. Secundus stakes a ducat and he will win the contest if he obtains two games before Primus obtains any. How much should Primus stake? It falls to Secundus to insist upon a sum such that the benefit that he will derive from a possible victory would be in an appropriate relation with the more unfavourable conditions which he accepted at the start. He must therefore define the terms of a new convention.

Secundus may reason as follows. Suppose that I effectively obtain two consecutive games.

(a) Therefore Primus will have to give me at least as much as if I had won two contests in playing according to the convention c_1 (*vincendo simpliciter 2 ludos*) that is 2 ducats.

(b) But that is not enough. For a player who won a contest according to convention c_1 keeps for himself [and] possesses in the strong sense the ducat which he won whereas I ran the risk of losing the second game after having won the first (*periculum perdendi secundum victo primo*). What I might have hoped for from this first success would then be wiped out. This possibility authorizes me to insist upon a compensation in the case of final victory: the ducat whose loss I risked after having won a game should be given to me if I actually win the second game.

Primus should therefore give me in total $2 + 1 = 3$ ducats. Secundus answers thereby to the very question asked: according to this new convention c_2,

---

[34] Ernest Coumet, 'Le problème de partis avant Pascal', *Archives internationales d'histoire des sciences, A 18* (1965), nos. 72/3, pp. 245–272, especially pp. 267–9.

The market place and games of chance in the fifteenth and sixteenth centuries 233

Primus who needs one game to win should stake 3 ducats. Secundus who needs to win two games staking 1 ducat.

3. New supposition: Secundus still stakes one ducat and must win three games before Primus wins one: what stake can he insist upon from Primus? We let him speak again:

Suppose that I have won three consecutive games. Primus should give me:

(a) At least as much as if I had won three contests played according to convention c_1; that is 3 ducats.

(b) A compensation that I can win in addition, for it was more difficult for me to win than for my adversary. To judge the amount of an analogous compensation, it was necessary earlier to think back to the moment when the penultimate game had not yet been played. Let us do the same here. Let us imagine that having won two consecutive games, I lose the third. This is the worst that can happen to me. How should the financial loss that I will then undergo be assessed? I declare that it is equivalent to the profit which I would have obtained from winning a contest played according to the convention c_2: in effect I ran the risk of seeing destroyed the earnings already acquired from two games. In supposing that these games were played according to an equitable convention, we know that they might bring me 3 ducats. These are therefore equally the 3 ducats which Primus should pay me as compensation for the peril of which I might have been the victim. The stake which Primus must commit, according to this convention c_3, will be of $3 + 3 = 6$ ducats.

In the general case we have a recursive method which states that the first has to stake c_n and the second 1 ducat in the case where the second needs n consecutive wins and the first only one. cC_n satisfies:

$$c_n = n \times c_1 + c_{n-1}.$$

Now Cardano has shown that

$c_1 = 1$, the *progressio* $^{1+1}C_2$ of 1,
$c_2 = 3 = {}^{2+1}C_2$, the *progressio* of 2, and
$c_3 = 6 = {}^{3+1}C_2$, the *progressio* of 3.

By the above condition it becomes clear that in general

$c_n = {}^{n+1}C_2$, the *progressio* of n.

So far Cardano has shown that if the first player lacks only one game for the total win and the second lacks n games, they have to share the stake in the ratio of

$${}^{n+1}C_2 : {}^{1+1}C_2 = {}^{n+1}C_2 : 1.$$

The last step in Cardano's approach is to extend this result to the case where the first player needs m games, where m can be different from 1. In this case the players would share the stakes in the ratio of

$^{n+1}C_2:^{m+1}C_2$.

The fascinating thing about Cardano's 'demonstration', or better explanation, is not, as has been stated in the literature, that this is 'one of the nearest misses in mathematics',[35] but that it is structurally similar to the triple contract, especially the first two parts of it. The similarity consists mainly in the role of risk. Risk and the difficulty of its evaluation is the point where Cardano departs from a unique mathematical solution to one that must appear only subjectively acceptable. This subjective element becomes even more explicit in a later treatment of the problem by Nicolò Tartaglia. Tartaglia accepts as reasonable the 'method of reduction' that Pacioli had rejected and says in this connection:[36]

> Therefore I say that such a problem can be solved better juridically than by reason; for no matter in what way one solves it there are always grounds for litigation.

No wonder that Tartaglia, in the light of this very fluid situation, claims only that his proposed solution is the least disputable.

The development following Tartaglia results from a change in the understanding of *fortuna* in the sense of chance as witnessed by Domingo de Soto's writings. This change permitted the clarification of what can be thought of as a 'just game'. Cardano made this step explicit in his later tract *De ludo aleae*. It is implicitly contained in his treatment of the problem of points in 1539, where he laid the foundation for the extension to more general forms of staking. Prior to Cardano, in the formulations of the problem of points each player had always staked the same amount with the implicit assumption that each player could win a single game as easily as his opponent. With Cardano the players can stake different amounts according to their different 'chances' of winning.

Forestani, a later critic of Pacioli, came on the scene[37] only in 1602. His treatment of the problem of points resembles the situation faced by Cardano and Tartaglia.

What has changed is that one can now openly discuss games of chance—still not in connection with the problem of stakes—and the role of *fortuna* or chance in making the outcome of an event unpredictable. Forestani as a member of the order *De minori conventuali di S. Francesco* could in 1602 openly speak of gaming situations that had been considered scandalous before, while the *Commissario generale* of his order gave his *imprimatur* on the grounds that he and the other censors had found nothing repugnant to Catholic faith or *buoni costumi*. This new liberality was prepared by the

---

[35] See M. G. Kendall, 'Studies in the history of probability and statistics. II. The beginning of a probability calculus', *Biometrika*, **43** (1956), pp. 1–14, especially p. 7 ff.

[36] Nicolò Tartaglia, *La prima parte del general trattato di numeri et misure*, Venice, 1556, fol. 165^v.

[37] See the first edition of the *Practica* . . . mentioned in (ref to note 4).

market place and the late sixteenth century's casuistic literature defending its practice.

The time was ripe to return to the mathematical standards that had been applied to the problem of points much earlier, as the Tuscan manuscript, written about 1400, demonstrates.

Perspective and the mathematicians: Alberti to Desargues

J. V. FIELD

IN mathematical terms, to construct a correct perspective picture is to carry out a form of conical projection. The centre of projection is the eye of the beholder (idealized as a point) and the projection is made by finding the points in which lines joining this point to points in the object or scene to be portrayed cut the surface upon which the perspective picture is to appear.

The name 'perspective' comes from the Latin word *perspectiva*, which was used in the Middle Ages to refer to the science of optics. This science included not only geometrical and physical optics but also physiological optics; that is, it was essentially a science of vision. To distinguish this 'perspective' from that used by artists it gradually came to be known as *perspectiva communis* while the science of perspective construction was called *perspectiva artificialis*.

The derivation of the name that is now applied only to the latter science does, indeed, seem to correspond fairly closely with the derivation of the science itself. For instance, in the earliest known written description of perspective construction, given by Leone Battista Alberti (1404–1472) in the first book of his *De pictura* (1435), the subject is introduced by a discussion of the mode of vision. Alberti refers to the 'cone of vision', formed by rays emitted from the eye (though he points out that the ensuing geometrical reasoning will also hold if the rays are those coming into the eye, as some natural philosophers prefer to believe is the case). The base of this 'cone' or 'pyramid' of vision is the outline of the object that is seen, so we are not dealing with a true cone in the mathematical sense. The picture surface, assumed flat, is described as a 'plane intersection' of the cone of vision. It is thus clear that Alberti is not thinking in terms of point by point projection from object to image. Indeed, it may be as well to remind ourselves that he would probably have taken the word 'projection' to refer to alchemists'

My essay sketches the history of how, in effect, mathematicians appropriated linear perspective and made it part of geometry. A complementary, and extremely interesting, study, which concentrates on the changing role of perspective in art, will be found in Martin Kemp's 'Geometrical perspective from Brunelleschi to Desargues: a pictorial means or an intellectual end?', *Proceedings of the British Academy*, 70 (1984), pp. 89–132, plus 13 plates (also issued separately ISBN 0 856 5250). I am grateful to Professor Kemp for very kindly allowing me to read his article before it was published. This general reference to it takes the place of the numerous specific ones I should otherwise have made in the notes that follow.

turning base metals into gold, though it was, apparently, also used by mathematicians to describe the method of constructing the flat images of the heavens found on astrolabes (the commonest observing instruments of the time).[1] In any case, Alberti's purpose in discussing the cone of vision, and proposing the picture as an 'intersection', is clearly to provide some background in conventional *perspectiva* for the linear perspective construction that follows.

Perspective for painters

Alberti's description of perspective construction is not notably clear and seems to be incomplete in various respects. For instance, it is not stated that the images of lines perpendicular to the picture plane will converge at the 'centric point' of the perspective (later to be known as the 'eye point' or the 'vanishing point'). However, there is general agreement that the construction he is describing is, essentially, that shown in Fig. 1.[2] This construction, or something very like it, was probably invented by Filippo Brunelleschi (1377–1446) in the second decade of the fifteenth century. It is well established that Brunelleschi made perspective pictures, the first of which had to be viewed through a peep-hole. Unfortunately, we have no direct evidence as to what led Brunelleschi to make these experiments or what method he used to construct the perspective of the pictures.[3] However, it seems certain that his early training as a goldsmith would have included quite a lot of mathematics of the 'practical' type found in abacus texts and it may also have included some of the astronomy and mathematics that would have been useful to a prospective maker of instruments. So it is conceivable that Brunelleschi was taught about the conical projection used to give a flat image of the sphere of the heavens on an astrolabe. We may note in passing that the only other artist who seems to have used perspective constructions as early as

---

[1] A connection between artificial perspective and the form of conical projection used for astrolabes was made explicitly by Federico Commandino in 1558, see below. Since astrolabes using this method of projection were well known throughout the Middle Ages and the Renaissance, it is very difficult to see why the rediscovery of Ptolemy's *Geographia* (which discusses conical projection) should be considered an important influence in the rediscovery of artificial perspective in the fifteenth century (as suggested in S.Y. Edgerton Jnr., *The Renaissance rediscovery of linear perspective*, New York 1975).

[2] The classical discussions of Alberti's construction are in E. Panofsky, 'Die Perspektive als "Symbolische Form" ', *Vorträge der Bibliothek Warburg* [1924-5], Berlin 1927, pp. 258-330, and W.M. Ivins, *On the rationalisation of sight*, New York (Metropolitan Museum of Art) 1938, reprinted New York (Da Capo) 1973. The standard modern edition of Alberti's work is *Leon Battista Alberti 'On painting' and 'On sculpture'*, ed. and trans. C. Grayson, London 1972. See also J. Gadol, *Leon Battista Alberti: Universal man of the early Renaissance*, Chicago and London 1969.

[3] An account of the available information and an incisive analysis of various scholars' interpretations of it are given in Martin Kemp, 'Science, non-science and nonsense: the interpretation of Brunelleschi's Perspective', *Art history,1* (1978), pp. 134-161.

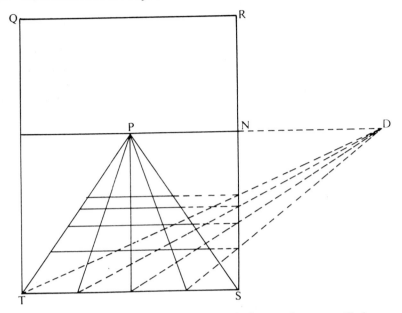

FIGURE 1. *Costruzione legittima*, to construct the image of a square-tiled pavement. P is the 'centric point' (also called the 'eye point', the 'vanishing point' or the 'principal vanishing point'). It is the foot of the perpendicular from the eye to the picture plane. Distance of eye from picture plane = DN.

Brunelleschi, namely his friend Donatello (1386–1466), was also trained as a goldsmith.

Donatello continued throughout his life to make use of perspective to give depth to his sculptured or cast reliefs.[4] There is no such straightforward and obvious connection between perspective and what seems to have been the main concern of Brunelleschi's life, namely his practice as an architect, though a very good case has been made out for the existence of a more subtle connection, in Brunelleschi's concern as to how the proportions between the parts of his buildings would appear from different positions.[5]

Despite the uncertainties surrounding Brunelleschi's perspective experiments, one must presume that he made some connection between his perspective construction and the 'cone of vision'. Alberti does not. Having discussed the 'cone' he passes on to describe the construction, merely asserting that it is mathematically correct (which it is). Alberti's lack of

---

[4] Sometimes to strange and powerful effect, as in the pulpits he made for the church of San Lorenzo in Florence in the last years of his life. See H. W. Janson, *The sculpture of Donatello*, Princeton 1963, p. 209 ff. for the pulpits.

[5] See R. Wittkower, 'Brunelleschi and "Proportion in perspective"', *Journal of the Warburg and Courtauld Institutes*, *16* (1953), pp. 275–291.

Perspective and the mathematicians: Alberti to Desargues 247

FIGURE 7. Piero della Francesca, *The story of the True Cross* (detail), probably painted about 1456–60 but designed somewhat earlier. The height of this register is 356 cm and the foreground figures are thus a little under life size. Photograph Alinari.

number of the examples correspond very closely with parts of Piero's fresco cycle *The story of the True Cross* in the church of San Francesco in Arezzo (Tuscany) and the beautiful illustrations which accompany them may well reflect the preliminary drawings Piero made in preparation for these paintings. We may note, nonetheless, that notwithstanding his known mathematical competence, and this evidence for its application to quite complicated elements in his pictures, Piero's actual use of perspective in this fresco cycle (and in his other pictures) is broadly similar to that of other painters. Like them, he uses buildings and other simple shapes to give a sense of space.[18] For example, in the scene from *The story of the True Cross* shown in Fig. 7, the foreshortened image of the cross serves to define the shape of the group of people round the young man who is being raised from the dead, while the surrounding architecture conveys the shape of the space within which the group is situated. The calculated foreshortenings of the heads seem to have the purpose of establishing that their perceived sizes and shapes will all be closely similar, giving a characteristic degree of uniformity to the figures.[19] Piero's use of perspective in his paintings, like his account of its use in his treatise, suggests that he felt no more need than other artists to explore the mathematics of the subject.

A mathematician's view in the 1580s

If Piero della Francesca's paintings present us with a non-resolvable mixture of mathematical and artistic skill (and why, indeed, should we expect the mixture to be resolvable?) his treatise on perspective in painting does, at least, treat almost exclusively of mathematics. It deals only with perspective, not with the multitude of other matters on which Alberti pronounced in *De pictura* (matters on which Piero also presumably had his opinions). Therefore, although the work is actually a practical manual addressed to painters, it is not unreasonable to regard *De prospectiva pingendi* as the first treatise on mathematical perspective.

This is in fact the view taken by a mathematician of a later generation, Egnazio Danti, in his preface to the perspective treatise by Vignola to which we have already referred: *Le Due regole della prospettiva pratica di M.*

---

[18] That he did so with mathematical precision, at least in one picture, has been proved by its having proved possible to reconstruct the ground plan of the scene in question—see R. Wittkower and B.A.R. Carter, 'The perspective of Piero della Francesca's "Flagellation"', *Journal of the Warburg and Courtauld Institutes*, 16 (1953), pp. 292–302. The ground plan is reproduced in Pirenne *op. cit.* in (note 12), above. There are some small departures from correct perspective, apparently dictated by Piero's concern for the surface geometry of his picture.

[19] Another characteristic of Piero's pictures, their extreme 'stillness', is connected with his concern for the surface geometry of his compositions. It has recently been suggested that, in one case at least, this concern took a highly sophisticated form, the composition being based upon a regular star pentagon—see B.A.R. Carter, 'A mathematical interpretation of Piero della Francesca's "Baptism of Christ"', in M.A. Lavin, *Piero della Francesca's 'Baptism of Christ'*, New Haven and London 1972, pp. 149–165.

FIGURE 8. Title page of Vignola, *Le Due regole della prospettiva pratica*, Rome 1583. Photograph courtesy of the British Library.

Iacomo Barozzi da Vignola . . . (Rome, 1583). (The title page of this work is shown in Fig. 8.) Danti adds that Piero's treatise is also 'the best in order and form' and comments on the excellence of its illustrations. This last remark must refer merely to the number and detailed nature of the illustrations since Danti says they can be judged from Daniele Barbaro's treatise[20]—in which the drawings are reproduced in the rather crude form of wood engravings. Danti's comments on Piero's treatise are part of a brief historical survey of writings on perspective, in the course of which he gives a representative list of the works concerned (all addressed in some measure to artists), together with succinct characterizations of their contents. Alberti, for instance, is described as merely giving a concise account of workshop rules; and the very popular treatise by Johannes Viator (Jean Pélerin), *De artificiali perspectiva* (Toul, 1505), is briskly (and not entirely unfairly) said to have 'a greater abundance of pictures than words'. Danti naturally concludes his survey by saying that none of the texts he has mentioned gives rules as sure and universal in their application as those in the treatise that follows—and the subsequent success of Vignola's work goes some way to vindicate this judgement.[21]

Egnazio Danti was a professional mathematician. He had been cosmographer to the Grand Duke Cosimo I of Tuscany and a professor of mathematics at the university of Bologna. In his preface to Vignola's treatise he writes very much as a mathematician, but we should not forget that he came from a family of practising artists[22] and must, from his earliest years, have been familiar with the manuals on perspective to be found in artists' studios. It appears from his brief history of such treatises that he had no very high opinion of most of them, as mathematical works. He clearly does not see any of them as contributing to the development of the mathematics of perspective. This seems, indeed, to be a fair assessment, though some of Danti's judgements on the earlier texts may have been coloured by his making insufficient allowance for the fact that the general level of mathematical competence was lower in, say, Alberti's day than in his own.[23] In contrast to Danti's approach, twentieth century attempts to descry mathematical 'progress' among the texts addressed to artists seem to me to be largely misguided.[24] All the texts are very elementary and their main purpose is not to

---

[20] D. Barbaro, *La pratica della perspettiva*, Venice, 1569, discussed below.

[21] Giacomo Barozzi da Vignola, *Le Due regole della prospettiva pratica*, edited and with commentaries by Egnazio Danti, Rome, 1583, preface by Danti. The work was reprinted several times.

[22] Its most famous member was Egnazio's brother, the sculptor Vincenzo Danti (1530–1576), whose large bronze group *The death of St John the Baptist* is still in position over one of the doors of the Baptistery of the Cathedral in Florence.

[23] I have considered Danti's history of perspective in more detail in 'Giovanni Battista Benedetti on the mathematics of linear perspective', *Journal of the Warburg and Courtauld Institutes*, *48* (1985), pp. 71–99.

[24] Traces of this approach are found, for instance, in R. Sinisgalli, *Per la storia della prospettiva (1405–1605). Il Contributo di Simon Stevin allo sviluppo scientifico della prospettiva artificiale ed i suoi precedenti storici*, Rome 1978.

give instruction in mathematics but to help the artist construct a picture. Viator's numerous illustrations may well have served instead of preliminary drawings, being transferred (if necessary in scaled form) to the panel that was to be painted.

The mathematization of the crafts

If Danti tends to ignore the purpose of the treatises he discusses, somewhat insensitively characterizing them merely as mathematically uninteresting, his list is nonetheless revealing in other respects. It shows rather well the gradual change in authorship of the treatises. The earliest are largely by painters: Piero della Francesca, Leonardo da Vinci (1452–1519), Albrecht Dürer. Later works are by architects, or show their influence, and there are also works by craftsmen, as well as one by a gentleman with an interest in mathematics, the Venetian humanist, diplomat and churchman Daniele Barbaro (1513–1570).[25]

It would, of course, be incorrect to describe Piero della Francesca or Leonardo da Vinci as being exclusively painters, but both had a much greater professional interest in painting than did most later writers on perspective, such as Vignola. What we seem to be seeing in this progress of perspective towards the applied arts in the sixteenth century is the progress of mathematics as an increasingly important component in the training and practice of craftsmen in general, and of architects in particular. For instance, mathematics came increasingly to be applied to such crafts as gunnery and fortification—as witness, for the former, the first mathematical treatise on ballistics: *La Nova Scientia* of Niccolò Tartaglia (1500–1557), published in Venice in 1537. Problems of fortification not only involved surveying (an art long recognized as connected with optics) but also the rather more arcane geometry required for designing the star-shaped forts that became fashionable in the sixteenth century as offering better defence against cannon.

The invention of such forts is generally credited to Pietro Cataneo (c. 1510–1569), who was by profession an architect, a military engineer and a teacher of mathematics.[26] The second edition of his treatise on architecture

[25] Barbaro was one of the founders of the Accademia Olimpica of Vicenza (1555). His wide-ranging learning is described in some detail in P. J. Laven, 'Daniele Barbaro, patriarch elect of Aquilea, with special reference to his circle of scholars and to his literary achievement', *unpublished Ph.D. thesis*, University of London 1957.

[26] For Cataneo's mathematical work, particularly as an algebraist, see R. Franci and L. Toti Rigatelli, 'La trattatistica matematica del Rinascimento senese', *Atti dell'Accademia delle Scienze di Siena detta de' Fisiocritici*, series XIV, *13* (1981), pp. 1–71, especially pp. 16–35. References to his work as an architect are given in Thieme-Becker *Künstler-Lexikon*. See also J. R. Hale, *Renaissance fortification. Art or engineering?*, London (Thames and Hudson) 1977.

The type of practical mathematical skill with which we are concerned here is paralleled by an increasing interest in more abstract considerations of mathematical proportions in architecture. Striking examples of such proportions are found in the work of Andrea Palladio (1508–1580), who was the architect of the Barbaro villa (at Maser) and, like Daniele Barbaro, a founder

(*Architettura*, Venice, 1567) contains a section on linear perspective which is mentioned by Danti in the history of perspective treatises to which we have already referred.[27] Military engineers were, of course, also concerned with *perspectiva communis*, not only because it supplied the theory underlying the use of sighting instruments for surveying, but also because geometrical optics supplied the answers to such important questions as how much of the gunner's view of an approaching enemy would be obstructed by a nearby bastion. An understanding of linear perspective, *perspectiva artificialis*, would have found practical application in, for instance, reconstructing the ground plan of a fort from a perspective sketch made by a scout, or in predicting how a proposed fort would appear to an approaching enemy.[28]

As mathematical knowledge was increasingly applied to the arts of war, mathematics naturally became a more important component in the education of the class that supplied the state with military commanders. It is therefore not surprising to find that Danti's list of perspective treatises should include one written by an amateur mathematician, the humanist Daniele Barbaro. Nor, indeed, is there any great distinction to be made between the levels of mathematical accomplishment that the practitioner, Cataneo, and the patrician, Barbaro, seem to expect in their readers.

Perspective for mathematicians, amateur and professional

Daniele Barbaro's treatise, *La Pratica della perspettiva* (Venice 1569), states on its title page that it will be very useful to painters, sculptors and architects. However, the intended readership of the work certainly includes the author's peers, that is gentlemen with an interest in mathematics, as well as (and probably in preference to) the artists whom they might employ. Thus Barbaro's mathematical treatise may be seen as complementary to, and *mutatis mutandis* addressed to much the same readership as, Alberti's *De pictura*, which had appeared in a new Italian translation in the previous year.[29]

---

member of the Accademia Olimpica. See R. Wittkower, *Architectural principles in the age of humanism*, London (Academy editions) 1973 (first published in 1949), and J.S. Ackermann, *Palladio*, London (Penguin) 1966.

[27] See note 21 above.

[28] See K. Veltman, 'Military surveying and topography: the practical dimension of Renaissance linear perspective', *Revista da universidade de Coimbra*, 27 (1979), pp. 329-368, for arguments for a long-standing connection between surveying and linear perspective (which Veltman does not, however, always sufficiently distinguish from *perspectiva* proper, that is Optics).

[29] L. B. Alberti, *Opusculi morali*, trans. and ed. Cosimo Bartoli, Venice, 1568. The translation of *De pictura* is dedicated to Giorgio Vasari. The translator, Cosimo Bartoli (1503-1572), had more than a passing interest in mathematics. He made an Italian version of Oronce Fine's *Protomathesis* (Paris 1532), published posthumously (Venice, 1587), and also wrote a work on surveying, *Del Modo di misurare le distanze* (Venice 1564). See J. H. Bryce, 'Cosimo Bartoli's *Del modo di misurare le distanze* (1564): A reappraisal of his sources', *Annali dell'Istituto e Museo di Storia della Scienza di Firenze*, V.2 (1980), pp. 19-34.

Although Barbaro's work is explicitly addressed to painters, the author complains in his preface that nowadays painters do not make much use of perspective. Naturalistic rendering of spatial relationships and relative sizes of figures and objects was, of course, a feature of the art of Barbaro's day— though not necessarily the dominant feature in most paintings. The use of perspective in this sense was, apparently, taken for granted, and at this time to say there was perspective in a picture meant neither more nor less than that it contained an architectural setting or vista.[30] In this sense, the title page of Vignola's treatise, shown in Fig. 8, provides a perfect example of perspective. Similar examples could be found in many fifteenth-century paintings, such as those shown in Figs 4, 5 and 7. However, the use of architecture in this manner was, indeed, not a marked feature of the painting of Barbaro's day, and particularly not in his native Venice. For instance, Titian (?1477–1576) uses architectural forms almost exclusively for structuring the surface geometry of his pictures (that is, as an element in their composition) rather than for providing a sense of depth. That is, he uses architecture as Piero uses the building in the background of the scene shown in Fig. 7 rather than in the way Piero uses the side buildings in the same picture. Jacopo Tintoretto (1518–1594), whose influence was second only to Titian's, shows a similar tendency to allow his figures to make their own space—though any general description of the style of his compositions would have to include some phrase such as 'expect the unexpected'. Tintoretto's picture *The origin of the Milky Way* (see Fig. 9),[31] probably painted in the late 1570s for the Emperor Rudolf II, provides a fairly extreme example of what Barbaro may have had in mind when he lamented painters' neglect of perspective.

On the other hand, perspective—in the narrow architectural sense—was a significant feature of contemporary stage scenery. Barbaro shows three examples of stage sets, all apparently taken from the *Libro di geometria e di prospettiva* (Paris, 1545) written by the architect Sebastiano Serlio (1475–1554). The sets correspond to those mentioned by Vitruvius in his *De architectura*, that is, they are for tragedy, comedy and satyr plays. Figure 10 shows Barbaro's version of tragic scenery, which, Vitruvius says, should show 'columns, pediments, statues and other objects suited to kings'.[32] Although this aristocratic architecture is much grander than that shown in the townscape that forms the background to the figures in Domenico Veneziano's *A miracle of St Zenobius*, shown in Fig. 2, the perspective

---

[30] T. Frangenberg, seminar at Warburg Institute, University of London, 1983. The substance of Frangenberg's argument on that occasion appears in his Ph.D. thesis (University of Cologne 1986).

[31] See C. Gould, *National Gallery Catalogues. The sixteenth-century schools*, London 1975, pp. 259–261.

[32] Vitruvius, *De architectura*, Book V, Chapter VI, see Vitruvius, *The Ten books on architecture*, trans. M. H. Morgan, New York (Dover) 1960 (first edition 1914), p. 150. Barbaro edited Vitruvius' text and wrote a long and learned commentary on it: *Marcus Vitruvius pollio de architectura libri decem, cum commentariis D. Barbari ... multis aedificiorum, horologiorum et machinarum descriptionibus, et figuris ... auctis et illustratis*, Venice, 1567; see also P. J. Laven *op. cit.* in (note 25) above.

FIGURE 9. Jacopo Tintoretto, *The origin of the Milky Way*, painted about 1576–80, 148 × 165 cm. (A piece has probably been cut away from the base of the picture. The missing part is believed to have shown a recumbent female figure representing Earth. The original height of the picture would have been about 260 cm.) Reproduced by courtesy of the Trustees, The National Gallery, London.

schemes are essentially similar. We have parallel lines of buildings perpendicular to the picture plane, various devices being used to break the monotony of the convergent lines and mitigate the sense of recession. Barbaro explains that the perspective construction for such stage scenery, consisting of a back piece and two pairs of side pieces, was carried out not geometrically but mechanically, using strings. His diagram of the procedure is not very clear, but it corresponds exactly with that shown in Fig. 11, which is taken from Vignola's perspective treatise of 1583.[33] As with the array of instrumental aids shown in earlier treatises, the mechanical method used to construct stage scenery shows that it was not considered necessary to search for a mathematical solution to the problems of perspective construction.

[33] P. 91 in *op. cit.* in note 21 above.

Perspective and the mathematicians: Alberti to Desargues 255

FIGURE 10. Stage scenery for tragedy, from Barbaro, *La Pratica della perspettiva*, Venice, 1569, p. 156 (after Serlio). Photograph courtesy of the Trustees of the Science Museum, London.

In fact, although Barbaro's treatise is much more elaborate than most of its predecessors—mainly because (as Danti remarked) it incorporates large portions of Piero della Francesca's treatise[34]—it nevertheless breaks no new ground as mathematics. It does, however, indicate the increasing importance given to mathematics as a component of a general education, for Piero's treatise had not seemed worth printing in earlier years, presumably because its mathematical level was relatively high.

Barbaro is certainly writing for a readership with an interest in mathematics but he is not concerned with the mathematics of perspective as such. In fact, he is consciously turning his back upon a truly mathematical treatment of perspective, for his preface refers to the difficulty of understanding the work on the subject by Federico Commandino (1509–1575), implying (misleadingly) that his own work contains a more elementary exposition of the same material.

[34] Barbaro acknowledges his debt to Piero. For Danti's comment see *op. cit.* in note 21 above.

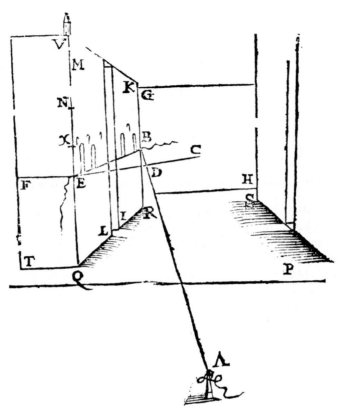

FIGURE 11. Constructing the perspective of stage sets by means of strings, from Vignola, *Le Due regole della prospettiva pratica*, Rome, 1583, p. 91. Photograph courtesy of the British Library.

Commandino, who made many excellent Latin translations of Greek mathematical works,[35] deals with perspective by way of preface to the discussion of stereographic projection that introduces his edition of the *Planisphaerium* of Claudius Ptolemy (fl. AD 129-141), published in Venice in 1558. Ptolemy's work requires such an introduction because it survives in a fragmentary and mathematically unsatisfactory state. However, despite its imperfections, its interest for Commandino and his contemporaries was not only as a work by the famous author of the *Almagest* but also as an authoritative Greek text on the mathematics required for astrolabes.[36] As we have

---

[35] See P. L. Rose, *The Italian Renaissance in mathematics*, Geneva (Droz) 1975.
[36] See R. B. Thomson, *Jordanus de Nemore and the mathematics of astrolabes: De plana spera*, Toronto 1978.

already noted, stereographic projection, which was the commonest form of projection used on astrolabes, is a form of conical projection and is thus mathematically equivalent to the projection used in making perspective pictures.[37] Commandino, writing as a mathematician for mathematicians, takes this equivalence for granted and proceeds to give a very short and straightforward account of perspective projection. He describes a number of special cases and one general one (for a general point), apparently taking the mathematician's somewhat impractical view that the problem is solved once one has dealt with a general case. This treatment would certainly have come as a shock to those accustomed to the litanies of worked examples found in manuals addressed to artists. However, Commandino does not show why the method of construction he describes is valid. Perhaps he thought that was obvious.

Commandino's brief remarks were taken up and greatly expanded by his pupil Guidobaldo del Monte (1545-1607), whose own treatise on perspective, *Perspectivae libri sex* (Pesaro, 1600), while largely derived from Commandino's work, is notable for giving the first general treatment of vanishing points—that is, showing that each set of parallel lines in the scene to be portrayed will become a set of lines convergent to its own particular 'vanishing point' in the picture.[38] Guidobaldo's treatise is also notable in that, following Commandino, it considers the projection of circles. Other treatises dealt only with rectilinear figures. For instance, Pietro Cataneo, in the section on perspective in the second edition of his *Architettura* (Venice, 1567) to which we have already referred, describes the perspective rendering of a regular polygon and then obtains the perspective version of a circle by joining up the vertices with a smooth curve—no doubt a reasonably easy exercise for an artistically talented reader. Guidobaldo, in contrast, recognizes that the projection of a circle will be another conic section, but his work essentially follows, though on a higher mathematical level, the pattern set by painters' manuals and here, as elsewhere, he makes no attempt to generalize. Thus each 'visual cone' contains a circle and one other conic section.

It was to be Girard Desargues (1591-1661) who first saw all the conics as sections of one cone, in the sense that they are all perspective images of one another. Moreover, given that this insight did not occur to either Commandino or Guidobaldo (of whose mathematical competence there can be no reasonable doubt) and given that Desargues in his work on conics deals first with properties of pencils, considering what properties are invariant under projection,[39] we may speculate that Desargues' invention of projective

---

[37] I have explained this more fully in the paper referred to in note 23 above.
[38] See B.A.R. Carter, 'Perspective', in H. Osborne (ed.), *The Oxford Companion to art*, Oxford (OUP) 1970, especially pp. 849-850.
[39] See R. Taton, *L'Oeuvre mathématique de G. Desargues*, Paris 1951. The work is translated in J.V. Field and J.J. Gray, *The geometrical work of Girard Desargues*, New York (Springer-Verlag) 1987.

geometry sprang not from an insight into the relation of conics to their cone but from the concept of invariance. We have thus arrived at the temptingly elegant thesis that Desargues' concerns mirror those of Brunelleschi. For considering invariance is the mathematical counterpart of what was probably Brunelleschi's (practical) reason for being interested in perspective, namely to know how it would affect the observer's apprehension of the proportions incorporated into the designs of buildings.[40]

It seems likely that Guidobaldo's interest in perspective was not only mathematical but also practical. One hesitates to describe him as a professional mathematician because he was a nobleman (the Marchese del Monte) and seems to have had no need to earn his living.[41] However, Guidobaldo practised as a military engineer and from 1588 was surveyor general of the fortifications of Tuscany. He is thus very much in line with the pattern of increasing mathematical skill found among surveyors, military engineers and architects.

Similarly practical concerns appear among the interests of another sixteenth-century mathematician who wrote on perspective, namely Giovanni Battista Benedetti (1530–1590). Benedetti seems to have written his brief treatise on perspective, *De rationibus operationum perspectivae*, while he was working in Turin as court mathematician to the Duke of Savoy. The final chapters of the treatise—which do not, however, have a necessary connection with the earlier part—are dedicated to the Duke's chief military engineer Giacomo Soldati (fl. 1561, died *c*. 1600). Benedetti's work was first published in 1585.[42] It is written for mathematicians and is utterly unlike any of its predecessors. Not only is it rigorous and elegant, but it also, as the title promises, explains the basis of perspective construction. The diagrams mainly come in pairs: one (marked *corporea*) to show the 'cone' or 'pyramid' of vision and the other (marked *superficialis*) to show the corresponding perspective construction as carried out in a drawing (see Fig. 12). Benedetti proves the equivalence of the diagrams in each pair. Strangely, his three-dimensional methods seem to have exercised no influence, though his work was apparently known to Simon Ṣtevin (1548–1620). Nevertheless, it is not merely as a beautiful piece of mathematics that Benedetti's work is of interest to the historian, for it provides striking evidence that by this time (the work was probably written in the 1570s) it was possible for a first-rate mathematician to interest himself in perspective as mathematics[43].

[40] See the paper by Wittkower referred to in note 5 above.
[41] This is well seen in the fact that, whereas most authors dedicate books to their patrons, Guidobaldo's are generally dedicated to members of his family.
[42] In a collection of numerous short works by Benedetti, some written many years before: *Diversarum speculationum mathematicarum et physicarum liber*, Turin, 1585. Second and third editions appeared in Venice, with slightly differing titles, in 1586 and 1599.
[43] A clear but very short account of Benedetti's treatise is given in H. Wieleitner, 'Benedetti als Perspektiviker', *Mitteilungen zur Geschichte der Medizin und der Naturwissenschaften*, 17 (Leipzig 1918, reprinted 1967), pp. 190–195. I have dealt with the work in some detail in the paper referred to in note 23 above.

Perspective and the mathematicians: Alberti to Desargues   259

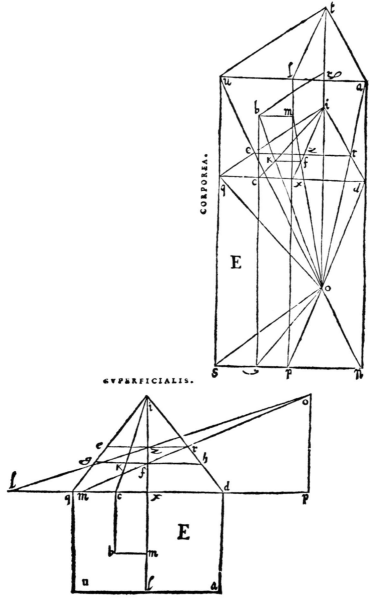

FIGURE 12. Figures in two dimensions ('superficialis') and in three ('corporea') to show the construction of the perspective image (*k*) of a general point (*b*) in the ground plane. The eye is at the point *o* and the picture plane is *qid*. From Benedetti, *Diversarum speculationum mathematicarum et physicarum liber*, Turin, 1585, p. 125. Photograph courtesy of the Trustees of the Science Museum, London.

The practical tradition: *perspectiva* and projective geometry

In the work of Commandino, Benedetti and Guidobaldo del Monte we see the theory behind the established craft of drawing in perspective being developed into an independent part of mathematics. That is, the mathematics becomes independent of the craft. It does not become independent of the branch of natural philosophy from which that craft was derived, namely the science of vision (which was in the later sixteenth century taking on the Greek title 'optics' in place of the Latin *perspectiva*). Indeed, the three-dimensional methods and diagrams in Benedetti's work are similar to those found in contemporary optical works such as Egnazio Danti's translation of Euclid, *La Prospettiva di Euclide* (Florence, 1573). Benedetti's diagrams are more complicated than those in optical works but it is clear that he is, essentially, regrafting *perspectiva artificialis* back onto its old stem of *perspectiva communis*.

As we have seen, geometrical optics had practical applications in surveying and fortification, and as the general mathematical competence required of practitioners of these crafts increased we find such practitioners not only capable of writing about perspective (as Cataneo in 1567) or discussing its mathematics with mathematicians (as Soldati in the 1570s) but also, in a later generation, making original contributions to mathematics—as did Simon Stevin, Girard Desargues and Albert Girard (1595-1632). The practical context of the study of perspective in the early part of the seventeenth century is well illustrated by the title page of the complete works of another military engineer, Samuel Marolois (dead by 1652), shown in Fig. 13. In the vignettes, perspective (top right) is associated with the crafts of the surveyor, the architect and the soldier, and the editor of the volume is the mathematician Albert Girard. It seems to be from this practical tradition, which preserved the close relations between natural and artificial perspective, rather than from its Renaissance adaptation for the use of artists, that Desargues' projective geometry derives. For our first hint of Desargues' new ideas comes in a postscript to his otherwise conventional, though characteristically short, text on perspective,[44] a postscript which draws a close analogy between the behaviour under the perspective projection of sets of parallel lines and lines which converge to a point—that is, ordinances with their butt at finite or infinite distance (as Desargues was to call them in his treatise on conics in 1639), pencils with their vertices at finite or infinite distance (as a modern

---

[44] [G. Desargues,] *Exemple de l'une des manieres universelles du S.G.D.L. touchant la pratique de la perspective sans employer aucun tiers point, de distance ny d'autre nature, qui soit hors du champ de l'ouvrage*, Paris 1636; reprinted exactly, apart from the odd misprint, in A. Bosse, *Maniere universelle de Mr Desargues, pour pratiquer la perspective par petit-pied, comme le geometral*, Paris 1648, pp. 321-334 and plate 150. Not reprinted in Taton *op. cit.* in (ref to note 39) above. Reprinted and translated in Field and Gray *op. cit.* in note 39 above.

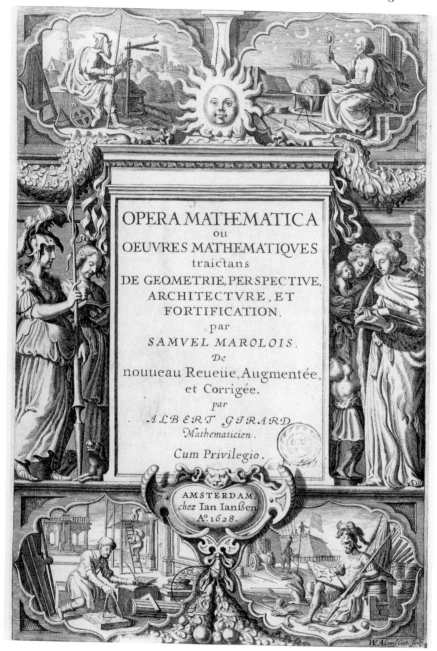

FIGURE 13. Title page of Marolois, *Opera mathematica*, Amsterdam 1628. Photograph courtesy of the Trustees of the Science Museum, London.

mathematician would say).[45]

In associating Desargues' invention of projective geometry with the practical tradition of optics, I do not wish to deny that Desargues' work, which was ultimately a contribution to pure geometry, must also be seen as having very important connections with the revival of interest in Greek geometry that took place in the Renaissance. As Jeremy Gray and I have argued elsewhere, Desargues must have known the relevant work of Euclid, Apollonius and Pappus;[46] and the exact nature and extent of their influence upon him is, naturally, of great interest to historians. In the present essay I have examined the perspective tradition, discussing what is, in effect, the social history of mathematics, because it seems that Desargues' work on projective geometry did also belong to the practical side of that tradition (as all his other work did) and took from it a crucial concern with three-dimensional problems, as in optics, rather than with the two-dimensional perspective constructions used by artists, or the essentially two-dimensional treatment of problems found in the tradition of Greek geometry.

There is, also, a further justification for supposing that the everyday concerns by which Desargues earned his living may have been relevant to the pure mathematical work which earns him a place in histories of mathematics. For projective geometry was invented not once but twice. After an initial period of interest, Desargues' work was not developed any further and was eventually forgotten. Projective geometry was invented again in the early nineteenth century. Again, its inventors were military engineers, pupils of Gaspard Monge (1746-1818) at the École Polytechnique in Paris, and trained by him in Descriptive Geometry (a technique for rendering three-dimensional shapes which has much in common with linear perspective). The parallel with Desargues' own practical concerns is as close as could be imagined between figures so far apart in time. It therefore does not seem rash to suggest that some of the factors which helped to determine the emergence of Desargues' projective geometry were connected with his professional concerns as a military engineer and architect.

The date of the emergence of projective geometry from the practical tradition of medieval *perspectiva* was determined by the progress of the mathematization of the crafts that had been hailed, close to its inception in

---

[45] See Field and Gray *op. cit.* in note 39 above. This concern with an invariant property tends to confirm the conjecture (expressed above) that in the train of thought that led Desargues to invent projective geometry the concept of invariance may have come before the idea of all the conics being projective images one of another. This property of the conics may, after all, be formulated as the proposition that the property of being a conic is invariant under the projection. See aslo J. V. Field, 'Linear prospective and the projective geometry of Girard Desargues', *Nunzio* (*Annali dell'Istituto e Museo di Storia della Scienza di Firenze*, new series) (in press).

[46] See Field and Gray *op. cit.* in note 39 above.

the case of painting, by Alberti's enthusiasm over the mathematical skills artists displayed in their use of linear perspective. Alberti undoubtedly overstated the claim of painting (even in his day) to be considered a mathematical art (on a par with, say, astronomy) but the phenomenon to which he drew attention was to prove of great historical significance. What he saw in the fine arts (as they later came to be known) was part of a renewal of vigour in the tradition of applied mathematics. The effect of that renewal on the history of natural philosophy was to be as profound as the effect on the art of succeeding centuries of the changes that took place in the art of fifteenth-century Florence.

Acknowledgements

I am grateful to Dr Jeremy Gray for making helpful comments on this article.

Why did mathematics begin to take off in the sixteenth century?

G. J. WHITROW

ONE of the great puzzles in the history of mathematics is why the subject as we know it today began to develop in the sixteenth century. This was before the main scientific revolution to which the sixteenth-century mathematical revolution can be looked upon as a kind of prelude. It is far from clear, however, why this earlier revolution came about. It is well known that the *De revolutionibus* of Copernicus, published in 1543, contained no new mathematical techniques. So, despite the fact that astronomy and mathematics were closely related subjects, the overthrow of Ptolemaic astronomy and the old cosmology was not initially dependent on new developments in mathematics as was the formulation of Newtonian astronomy and natural philosophy. But these later developments in mathematics were themselves dependent on the principal advances that occurred in the sixteenth century.

During the Middle Ages the status and reputation of mathematics and mathematicians were ambiguous. Roger Bacon in 1267 complained of writers who counted mathematics among the seven 'black arts', and this belief survived in the sixteenth century and later. It is notorious that religious reformers in the time of Edward VI destroyed mathematical manuscripts at Oxford under the delusion that they were diabolical, and this may account for the disappearance at that time of many of the works of the fourteenth-century school of astronomers.[1] Even in the seventeenth century John Aubrey mentioned that, following the foundation of the Savilian chair at Oxford, some parents were alleged to have kept their sons from that university fearing that they might be 'smutted with the black art'.[2] On the other hand, throughout the Middle Ages tribute was repeatedly paid to the power and importance of mathematics by men of learning, not least by Roger Bacon. Nevertheless, progress was very slow. Much time and energy were devoted to the retrieval and preservation of ancient mathematics, including the works of Archimedes, and some new investigations were made as a result, particularly in extending the Greek theory of proportion in a way which foreshadowed the modern concept of functional variation. Subtle

---

[1] R. T. Gunther. *Early science in Oxford*. Oxford, 1930. Vol. 2, pp. 42–3.
[2] E. G. R. Taylor. *The mathematical practitioners of Tudor and Stuart England*. Cambridge, 1954, p. 8.

consideration was given to infinite series and the continuum by scholastic philosophers, but much of it was verbal rather than mathematical. Duhem has argued that the comparative absence of mathematical skills in their work may have been largely due to the craze for dialectic and the desire to win victory in debate rather than to search for the truth.[3]

A curious feature of the Middle Ages that persisted throughout the sixteenth century was that the general attitude to novelty tended to be negative. For, although as we now realize there was a great deal of technological progress in Western Europe in the Middle Ages, its psychological effects seem to have been slight and it generated no general concept of technical progress. 'An inventor', it has been said, 'was . . . a person who found something which had been lost, not one who devised a new solution unknown to previous generations.'[4] Moreover, the spirit of humanism and the literary renaissance was backward-looking. The extraordinary thing is the way in which some of those responsible for important new developments were influenced by this. A remarkable example was the key figure in the development of algebra, Vieta, for his humanistic education led him to regard innovation as being essentially renovation. Nevertheless, during the course of the sixteenth century men here and there began to rebel against the tyranny of antiquity. A contemporary of Vieta who was influential in the spread of the use of decimal fractions, Stevin of Bruges, had little respect for the authority of the Greeks and instead was always drawing on his practical experience.

One of the factors that limited the communication of new ideas and made it more difficult for mathematics and science to be the collective collaborative disciplines that they are today was the cult of secrecy, encouraged by the fear of plagiarism. The invention of printing with movable type in the middle of the fifteenth century was eventually responsible for overcoming this particular obstacle to the dissemination of knowledge, but how important was this invention for the advance of mathematics in the sixteenth century? Elizabeth Eisenstein has complained that, when historians consider the various factors that may have influenced the great scientific innovators of the sixteenth and seventeenth centuries, there has been a tendency to pay too little attention to the fifteenth-century communications revolution. She maintains that

If communications are considered in some detail, or the shift from script to print is assigned importance, the discussion is more likely to focus on seventeenth-century correspondence than on fifteenth-century books.[5]

---

[3] P. Duhem. *Le système du monde*. Paris, 1957. Vol. 7, pp. 621–2.
[4] K. Thomas. *Religion and the decline of magic*. London, 1971, p. 430.
[5] E. Eisenstein. *The printing press as an agent of change*. Cambridge, 1980, p. 460.

The printed word first assumed importance for the European scientific community when Regiomontanus established his Nuremberg press. The most influential mathematician of the fifteenth century, and one of the few of his day who knew Greek, he set up a printing press in his own house about 1470. He was the first publisher of astronomical and mathematical literature and he sought to assist scientists and mathematicians by providing them with texts including the works of Archimedes, Apollonius and Ptolemy, free of scribal and typographical errors, but unfortunately he died in 1476. He himself composed the earliest systematic treatise on trigonometry, which he was the first to organize as a mathematical discipline independent of astronomy. His *De triangulis*, however, did not appear in print until 1533. The invention of printing would, of course, have made far less impact if it had not been possible to manufacture paper at a reasonable price. Another important technological development was type-cutting. During the sixteenth century 'roman' and 'italic' type gained over 'gothic', except in Germany. This was partly because they were more flexible, particularly in their capacity to combine upper- and lower-case letters. This was not without effect on the development of mathematical symbolism.

The choice of books printed was, however, influenced by economic factors, that is to say by demand. A printer who failed to realize this soon went bankrupt. As a result, the general effect of printing was less immediate than it might have been because the knowledge first transmitted by means of it tended to be traditional. Moreover, unlike Regiomontanus, some of the prominent scientists and others were slow to appreciate the significance of printing. For example, although Leonardo da Vinci invented and reinvented many different machines, he overlooked two of the greatest inventions of his day: printing and engraving.[6] Not only did Copernicus not rush into print, but he was unfamiliar with Regiomontanus's *De triangulis* until 1539, when Rheticus brought him a copy. As Elizabeth Eisenstein has remarked

That Copernicus had already finished his section on trigonometry without any knowledge of his predecessor's treatise suggests how scientific interchange was enfeebled when texts were left in manuscript form. How Regiomontanus's influence was extended by the use of the Nuremberg press is indicated by the fact that Copernicus overhauled his presentation of indispensable theorems after consulting the fifty-year-old work.[7]

Once his work did get printed, Copernicus's views were, however, slow to spread. Printers were not mobilized by astronomers and mathematicians as they were by religious reformers. Moreover, it was the *Sphaeri Mundi* of Sacrobosco and not *De revolutionibus*, or for that matter Ptolemy's

---

[6] G. Sarton. *Six wings*. Bloomington, Indiana, 1957, pp. 228–9.
[7] E. Eisenstein. *op. cit.*, p. 591.

Almagest, that ran through two hundred editions by 1600, of which at least thirty appeared between 1472 and 1501.

The key subject in the mathematical revolution of the sixteenth century was algebra, and the first book on the subject that was printed was the *Summa de arithmetica* of Luca Pacioli, which appeared in 1494. It was extremely influential, presumably because it was printed. The earliest Renaissance algebra was that of Chuquet, but although it was not without effect this was no doubt less than it might have been because it was not printed. Neither Pacioli nor Chuquet, however, was the first algebraist of importance in Western Europe. The first was Fibonacci (Leonardo of Pisa), who flourished early in the thirteenth century. His misnamed *Liber abaci* was important in the transmission of the Hindu-Arabic numerals and system of numeration, although these were slow to be adopted. Moreover, as Boyer has pointed out,

It is one of the ironies of history that the chief advantage of positional notation—its applicability to fractions—almost entirely escaped the users of the Hindu-Arabic numerals for the first thousand years of their existence.[8]

In this respect, Fibonacci was as much to blame as anyone, and as a result the general use of decimal fractions was delayed for another three hundred years until Stevin's advocacy of their use. Another able mathematician of the thirteenth century was Jordanus Nemorarius who went further than Fibonacci in his use of letters. He often used a single letter to denote a number, although he sometimes used two letters. The significance of Jordanus in the history of algebra has been stressed by Barnabas Hughes, the editor and translator of Jordanus's treatise *De numeris datis*. Hughes makes the important point that Jordanus's method of dealing with equations avoided the traditional appeal to geometry, an equation being subjected to transformations until it acquired a canonical form which immediately provided an algebraic solution.[9] Jordanus stated the solution of a quadratic in general form, but when applying it to particular examples he used Roman numerals!

The same century saw the translation by William of Moerbeke of the main works of Archimedes from Greek into Latin, but this translation had only a modest effect in the next two centuries, although in the Renaissance it exerted a considerable influence, being the first to be printed early in the sixteenth century.[10] Meanwhile, in the fourteenth century there were important advances in the concepts of function and the continuum and Oresme introduced the device of representing a function graphically, a step towards

---

[8] C. B. Boyer. *A history of mathematics*. New York, 1968, p. 280.

[9] Jordanus de Nemore. *De numeris datis*. A critical edition and translation by Barnabas Bernard Hughes. Berkeley, Cal., 1981, p. 14.

[10] M. Clagett. 'The impact of Archimedes on medieval science'. *Isis*, 50 (1959), p. 428.

the later development of analytical geometry. He also derived rules for combining proportions that were equivalent to our laws of exponents. John Murdoch has pointed out that

the most important distinguishing mark which the medieval history of proportions exhibits is its consistent tendency to read arithmetical conceptions into the geometrical, and into theories dealing with general magnitude. In effect, number was being considered an element of geometry: the Greek distinction between the continuous and the discrete was beginning to undergo erosion.[11]

In short, arithmetic was in the ascendant.

All these promising beginnings in the thirteenth and fourteenth centuries tended, however, to peter out and arithmetical algebra languished until Pacioli found a manuscript of Fibonacci's *Liber abaci* in the library of San Antonio di Castello at Venice.[12] According to Cardano, Pacioli openly revived and incorporated Fibonacci's ideas in his encyclopedic *Summa de arithmetica* of 1494, ideas which were subsequently developed by Scipione del Ferro, Tartaglia and Cardano himself. In his book Pacioli mentioned that a method of solving the cubic equation had not yet been found. Whether it was this that stimulated Scipione (a professor of mathematics at Bologna) to devise a successful method, soon after 1500, is uncertain, but to us today it seems very strange that over forty years elapsed before the method was published, in 1545 in the *Ars magna* of Cardano. He stated explicitly that the solution was not his, but he clearly thought that it was high time that it was published. Meanwhile other textbooks on arithmetical algebra had appeared in Germany, but they were eventually overshadowed by Cardano's book, which not only gave the solution of the cubic equation but also the solution of the quartic by Cardano's pupil Ferrari. It was a book that was not in the Greek tradition at all but in that of al-Khwarizmi. It contained little symbolism and only specific numerical coefficients. Nevertheless, its publication was an important step towards regarding algebra as a mechanism for combining symbols without concentrating on their geometrical or physical meaning. Moreover, it directed attention towards extensions of the concept of number, including complex numbers which figured in Bombelli's *Algebra*, published in 1572. The latter was the most systematic treatise on algebra produced in Italy in the sixteenth century and was intended to replace Cardano's *Ars magna*. Influenced by Diophantus, Bombelli thought, however, that he had restored the effectiveness of arithmetic by imitating the ancient writers.[13]

With Bombelli the sixteenth-century Italian development of algebra came

---

[11] J. E. Murdoch. 'The medieval language of proportions', in *Scientific change* (ed. A. C. Crombie). London, 1963, p. 270.
[12] P. L. Rose. *The Italian renaissance of mathematics*. Geneva, 1975, p. 83.
[13] P. L. Rose. *op. cit.*, p. 147.

to an end. Why then did its influence not peter out in the way in which that of the medieval mathematical developments, to which I have already referred, did? There were no further developments in solving equations of higher degree, algebraists being in fact faced with insoluble problems—a situation broadly similar to that associated in antiquity with the three classical geometrical problems of Plato—and not until the time of Galois were important advances made. Indeed, in Italy at the beginning of the seventeenth century there was no sign that any major new achievements in algebra were likely to occur in the foreseeable future. Instead, Galileo in developing his 'two new sciences' remained a faithful follower of Archimedes and Euclid and made no use of the algebraic techniques available in his day. This is all the more striking in view of his fascination with infinitesimals which brough him close to discovering that there are infinitesimals of different order.

It was in France, rather than in Italy, that the sixteenth-century reaction against Greek geometrical techniques was decisive for the future of mathematics. The great power of 'algebra', in the general sense that includes 'analysis', only began to emerge with Vieta's partial, but important, development of symbolism (which distinguished for the first time between a parameter and an unknown quantity) and particularly by his new generalized concept of number. Vieta has been described as 'the first mathematician of his age to think occasionally as mathematicians habitually think today.'[14] He was the true founder of algebraic-style analysis, although, as I have already mentioned, he was, paradoxically, like so many of his contemporaries, one who regarded 'innovation' as 'renovation'. Jacob Klein has said that the founders of modern science 'were not, for the most part, aware of their own conceptual presuppositions.'[15] Nevertheless, Vieta was aware that his algebra was a study of general forms and expressions and was more than symbolic arithmetic. His *Introduction to the analytical art* ends with the famous claim that this art 'appropriates to itself by right the proud problem of problems, which is: *to leave no problem unsolved.*' Thus 'algebra' was set on the path by which it eventually supplanted geometry as the fundamental branch of mathematics and the most powerful mathematical tool, but why sixteenth-century mathematics should have had this remarkable outcome instead of petering out is far from clear.

[14] E. T. Bell. *The development of mathematics.* New York, 1940, p. 99.
[15] J. Klein. *Greek mathematical thought and the origin of algebra* (translated E. Brann). Cambridge, MA., 1968, p. 152.

INDEX

abbaci 134
abbacists 21-2, 27
abbacus tradition 130, 132-3
  French 135-7, 144
abbacus treatises 15
Adam, Jehan 65, 135-6
Agrippa, Henry Cornelius 7, 209-19
Alberti, Leone Battista 236-9, 243, 262
Albert of Saxony 42-6, 48
Alcuin 67, 204-5
Alexander 72
Alexander of Villedieu 74, 132
algebra and business problems 11, 27-8
algebra
  development 11-29, 188-9
  Italian, sixteenth century 11
algebraic notation 3-4
algebraic rules 16, 18
algebraic terms, Italian 15-16
algorisms 134
al-Khwarizmi 18, 79, 127-8
Allard, André 86
Al-Nasawi 90-1, 95
al-Uqlidis 51
Aquinas, Thomas 222-3
Archimedes 62, 267
Aristotle 40-1

Bachet de Méziriac 162-3
Bacon, Roger 264
Baker, Humphrey 124
Baldi, Bernardino 2, 190-3
Barbaro, Daniele, Vignola, G.B. da 252-5
Bellini, Giovanni 244
Benedetti, Giovanni Battista 258-60
Benedetto, M° 14, 27
Berselius 76
Biagio, M° 27
Boethius 62, 165
Bombelli, Rafael 72, 124-6, 143, 268
Boncompagni, Baldassare 190-4
Borghi, Pietro 111
Borrel, Jean (Johannes Buteo) 140
Bouvelles, Charles de 139
bracketing 121
Brunelleschi, Filippo 237-8, 244
Bullettino di bibliografia e di storia della matematiche e fisiche 7, 59, 66, 190-4

Burckhardt, Jacob 2
Buridan, John 43, 132
Buteo, Johannes 38, 52, 121, 123-4, 141

Campanus 62
Cantor, Georg 41
Cantor, Moritz 124
Cardano, Girolamo 5, 11, 52, 92, 94-5, 124-5, 129, 132, 141, 221, 231-4, 268
Cataldi 50
Cataneo, Pietro 251
Certain, Jehan 78, 81
Ceulen, Ludolf van 160
Chasles, Michel 61, 124-5
Chiarini 118
Chuquet, Nicolas 3, 5, 38, 59-126, 135-6, 138, 140, 186-90
  Triparty 35
Clagett, Marshall 132
coinage problems 105-6
Commandino, Federico 255-7, 260
compound interest 107
Copernicus 264, 266
cube roots 89

da Firenze, Antonio 13
Dagomiri, Paolo 98
Damhouderius, Jodocus 226-7
Danti, Egnazio 248, 250-1
da Perugia, Masolo 22-3
da Vinci, Leonardo 251
de' Bicci de' Medici, Giovanni 22-3
de' Danti, Giovanni 13
de Guibert, G. 131
de la Roche, Estienne 6, 31, 38, 61, 67, 72, 87-8, 117-21, 123-6, 139-40
del Ferro, Scipione 11, 129, 132
dell'Abbaco, Paolo 27, 98
della Francesca, Piero della 6, 8, 246, 248, 250-1, 255
del Monte, Guidobaldo 257-8, 260
de' Mazzinghi, Antonio 15, 17, 28
de Meré 221
de Nemore, Jordanus 4, 186, 267
de Ortega, Juan 38
de Roman(i)s, Berthelemy 62, 68, 87, 112, 135
Desargues, Girard 257-8, 260, 262

Descartes, René 126-7
de Soto, Domingo 226, 234
Diophantus 126, 143, 162
Donatello 238, 244
Dürer, Albrecht 241, 251
Dutch Mathematical School 160

Egmond, Warren van 77
Eisenstein, Elizabeth 1, 265-6
equations 11, 18-27
Euclid 70
Euler 68
Eutocius 70
Eves, Howard 117

Favaro, Antonio 192-3
Fermat, Pierre de 7, 126, 162-3
Ferrari, Ludovico 129
Fine, Oronce 139-40
Forestani 234
Franci, Raffaella 119, 129
Frescobaldi, Philippe 87-8, 119
Frisius, Gemma 148
Fulconis, Johan Frances 30-1, 124
Fusoris, Jean 74-6, 80, 135

Galilei, Galileo 41
geometry, projective 262
Gerardi, Paolo 12, 27, 82
Gerard of Cremona 70, 128
Ghaligai 119
Gilio 13
Gori, Dionigi 138
Gosselin, Guillaume 124, 140, 142

Hebrew tradition of mathematics 137
Heron of Alexandria 90-1, 93-5
Hilbert, David 43
Hispalensis, J. 49-50
Hughes, Barnabas 186, 267

induction (in Maurolico) 171, 176
Italian mathematical manuscripts 8, 11-29
Itard, Jean 59, 64, 96

Jean de Murs 77-8, 90, 131
John of Seville 64
Juschkewitsch, A. P. 119-20
Justus, Pascasius 225-6

Kadran aux marchans 78, 81-2, 96, 108, 135

Klein, Jacob 269
Kushyar Ibn Labban 90-1

language, scientific 184
Lefevre d'Etaples, Jacques 132, 139
Leonardo of Pisa 4, 11, 18, 29, 31, 49, 52, 62, 70, 78, 90, 92, 95, 118, 267
  Liber abaci 50, 77, 128, 130-1
Levey, Martin 89
L'Huillier, H. 59-60, 84, 87, 97, 187
Libri, Count 2
logarithms 63-4
Lull, Ramon 62, 70
Lyon 71

maestri d'abbaco 128-9
magic squares 214-17
Magnus, Albertus 218
Manuscrit français 2050 80-2, 112
Marolais, Samuel 260-2
Marre, A. 59-61, 66, 78, 82, 96, 112, 117
mathematical magic 209-19
mathematics
  development of 8
  recreational 7, 195-208
Matthew effect 4
Maurits, Prince of Orange 158
Maurolico, Francisco 2, 7, 162-79
Mazzinghi, Antonio de 227
Merton, Robert 4
Metius, Adriaan 160
Metrodorus 66
Moerbeke 70
Monge, Gaspard 262
Murdoch, John 268

Narducci, Enrico 193
Needham, Joseph 185
Nicomachus of Gerasa 165
number (Agrippa's concept) 211-13
numbers
  central regular polyhedral 169
  'column' 169
  figurate 166
  figurate (table) 172
  'heteromecic' 166
  linear 166
  negative 63
  perfect 170
  polygonal 168
  pyramidal 168
  solid 166
  superficial 166-7
  surface 166

numeration 65, 121
Nunez, Pedro 143

occult virtues 217-19
Olivi, Peter John 223
Omar Khayyam 127-8
Oresme, Nicole 42, 62, 132, 267

Pacioli 5-6, 11, 31, 65, 70, 72, 84, 98, 107, 118-21, 141, 220-1, 230-1, 246, 267-8
Pamiers manuscript 5-6, 30-1, 79, 82, 85, 135
Pappus 37, 51, 126
Pascal, Blaise 7, 162, 221, 225
Pastoureau, Michel 77
Peletier, Jacques 125, 140-1
Pellos, Frances 30, 78, 81-2, 84
  Compendion 135
per chasella rule 16-17
Perez de Moya, Juan 38
perspective 8
perspective, artistic 236, 260, 262-3
perspective, mathematical theory 8, 236, 260, 262-3
Poulle, Emmanuel 74
Prehoude, Matthew 135-6
printing 265-7
priority claims 5
probability theory 7-8, 220-5
problem
  cistern 198-200
  hundred fowls 202-3
  knight's tours 201
  monkey and coconuts 201-2
  of points 7-8, 220, 235
  river crossing 204
  transport 204-5
Psellus, Michael 64

Ramus, Petrus 132, 140, 142, 157-8
Rashed, R. 73
reasoning, recurrent (in Maurolico) 173-5
Regiomontanus 2, 266
Renaissance 1-2
Renaissance and mathematics 1
Robert of Chester 128
Roland l'Ecrivain 78
Rolland of Lisbon 131
roots
  bound 121
  cube 89-94
  imperfect 32-6, 38, 40, 48-53, 85-6
Rose, Paul Lawrence 2
Rostow, W. W. 8
rule of first terms 65, 72, 187
rule of first terms (canons) 123
rule of intermediate numbers 5, 36-7, 62, 65-6, 72, 84, 86, 121
rule of three 99-100, 103, 110-11

Sacrobosco 74, 80, 112-13, 132
San Bernardino of Siena 223
Sanct Climent, Francesch 78-9
Sarton, George 1, 137
Scheubel, Johann 140
Schooten, Frans van 156, 161
Sédillot 192
Sesiano, Jacques 79, 81, 85, 135
Snellius, Rudolf 6-7, 156-8, 160-1
Steiner, George 183
Steinschneider 192
Stevin, Simon 72, 140, 143, 160
Stifel, Michael 39, 125, 141
Sullivan, J. W. N. 117

Tartaglia, Niccolo 52, 129, 143, 234, 251
Thorndike, Lynn 1
Tintoretto, Jacopo 253-4
Toti Rigatelli, Laura 129, 221, 227
translation 183-9
triple contract 224-5

Uccello, Paolo 240

Vacca, G. 163
Vanden Hoecke, Gielis 147, 149
Vander Noot, Thomas, *Die Maniere* 147, 149-51, 153
Varenbraken, Christianus van 147-50, 154
Vasari, Giorgio 3, 6, 240
Veneziano, Domenico 243, 253
Viéte, François 6, 125-7, 143, 265, 269
Villani, Giovanni 97

Xylander, G. 143, 162